DEVELOPING FRACTIONS KNOWLEDGE

SAGE was founded in 1965 by Sara Miller McCune to support the dissemination of usable knowledge by publishing innovative and high-quality research and teaching content. Today, we publish over 900 journals, including those of more than 400 learned societies, more than 800 new books per year, and a growing range of library products including archives, data, case studies, reports, and video. SAGE remains majority-owned by our founder, and after Sara's lifetime will become owned by a charitable trust that secures our continued independence.

Los Angeles | London | New Delhi | Singapore | Washington DC | Melbourne

DEVELOPING FRACTIONS KNOWLEDGE

AMY J. HACKENBERG
ANDERSON NORTON
ROBERT J. WRIGHT

Los Angeles | London | New Delhi
Singapore | Washington DC | Melbourne

Los Angeles | London | New Delhi
Singapore | Washington DC | Melbourne

SAGE Publications Ltd
1 Oliver's Yard
55 City Road
London EC1Y 1SP

SAGE Publications Inc.
2455 Teller Road
Thousand Oaks, California 91320

SAGE Publications India Pvt Ltd
B 1/I 1 Mohan Cooperative Industrial Area
Mathura Road
New Delhi 110 044

SAGE Publications Asia-Pacific Pte Ltd
3 Church Street
#10-04 Samsung Hub
Singapore 049483

Editor: Marianne Lagrange
Editorial assistant: Robert Patterson
Production editor: Tom Bedford
Copyeditor: Elaine Leek
Indexer: David Rudeforth
Marketing manager: Dilhara Attygalle
Cover design: Naomi Robinson
Typeset by: C&M Digitals (P) Ltd, Chennai, India
Printed and bound in Great Britain by Ashford
Colour Press Ltd.

Library of Congress Control Number: 2015960206

British Library Cataloguing in Publication data

A catalogue record for this book is available from
the British Library

ISBN 978-1-4129-6219-3
ISBN 978-1-4129-6220-9 (pbk)

To Becky, Caroline, Eleanor, and Nikolas

Brief Contents

Contents

List of Figures

List of Tables

About the Authors

Dr Amy J. Hackenberg holds a Bachelor's degree in visual and environmental studies from Harvard University, in which she studied architecture in an attempt to unite the arts and mathematics. Following an initial mathematics teaching job in which she taught middle and high school students in Los Angeles, she earned a Master of Arts in Teaching degree in mathematics education at the University of Chicago, and after more teaching of high school students, she earned a doctoral degree in mathematics education from the University of Georgia. Amy is an associate professor of mathematics education at Indiana University–Bloomington. For the last ten years she has conducted research on how middle school students construct fractions knowledge, on relationships between middle school students' fractions knowledge and algebraic reasoning, and on how teachers can learn to develop student–teacher relationships that meet the cognitive and emotional needs of students and teachers. In her current project she is investigating how to differentiate instruction for cognitively diverse middle school students (see www.indiana.edu/~idream), studying her own teaching as well as working with practising teachers.

Dr Anderson Norton is associate professor of mathematics education in the Department of Mathematics at Virginia Tech – his alma mater. Having completed a Bachelor of Science degree in mathematics from Virginia Tech, and following a brief career as an engineer, he pursued graduate studies in mathematics and mathematics education at the University of Georgia. In the middle of his graduate years, he taught high school mathematics at Loganville High School and at the Governor's Honors Programme in Georgia. He holds a Master of Science degree in mathematics and a Doctor of Philosophy degree in mathematics education. His research focuses on building psychological models of students' mathematical development, especially in the domain of fractions. More broadly, he is interested in the nature of mathematics and mathematical learning. For his efforts to build bridges between teaching and research, he was awarded the 2013 Early Career Award from the Association of Mathematics Teacher Educators. These efforts include the design of apps that support students' development of fractions knowledge (see http://ltrg.centers.vt.edu).

Dr Robert J. (Bob) Wright holds Bachelor's and Master's degrees in mathematics from the University of Queensland (Australia) and a doctoral degree in mathematics education from the University of Georgia. He is an adjunct professor in mathematics education at Southern Cross University in New South Wales. Bob is an internationally recognized leader in assessment and instruction relating to children's early arithmetical knowledge and strategies, and is the lead author of four books, and many articles and papers in this field. His work over the last twenty years has included the development of the Mathematics Recovery Programme which focuses on providing specialist training for teachers to advance the numeracy levels of young children assessed as low attainers. In Australia and New Zealand, Ireland, the UK, the USA, Canada, Mexico and elsewhere, this programme has been implemented widely and applied extensively to classroom teaching and to average and able learners as well as low attainers. Bob has conducted several research projects funded by the Australian Research Council, including the most recent project focusing on assessment and intervention in the early arithmetical learning of low-attaining 8–10 year olds.

Acknowledgments

This book offers testimony to the legacy of Les Steffe and the research programme he began with Ernst von Glasersfeld in the 1980s – a research programme that fully recognizes students as mathematicians and fully honours the mathematics that they construct. Les investigates students' mathematics because he finds their worlds fascinating, and he introduced each of the authors to the creative endeavour of building models of those worlds. This book is a product of that work.

The mathematical ideas presented here, we learned from students. We thank those students for the opportunity to study fractions knowledge and its development. Specifically, we recognize the students, teachers and administrators in the following schools: West Jackson Middle School in Georgia; Blacksburg Middle School and Shawsville Middle School in Virginia; and Tri-Noth Middle School in Indiana, particularly teachers Mark Eckerle, Darrell Frazier, Patti Walsh, and Principal Craig Fisher. We also recognize the organizations that have provided support for our work: the United States Math Recovery Council; Indiana University, through a Proffitt grant; Virginia Tech's Institute for Society, Culture and Environment; and the National Science Foundation, through a pair of grants (DRL 1252575 and DRL 1118571).

Series Page

This book – *Developing Fractions Knowledge* – is a significant and important addition to the current Mathematics Recovery Series. The five books in this series address the teaching of early number and whole number arithmetic and fractions in primary, elementary and secondary education. These books provide practical help to enable schools and teachers to give equal status to numeracy intervention and classroom instruction. The authors are internationally recognized as leaders in this field and draw on considerable practical experience of delivering professional learning programmes, training courses and materials.

The books are:

Early Numeracy: Assessment for Teaching and Intervention, 2nd edition, Robert J. Wright, Jim Martland and Ann K. Stafford, 2006.

Early Numeracy demonstrates how to assess students' mathematical knowledge, skills and strategies in addition, subtraction, multiplication and division.

Teaching Number: Advancing Children's Skills and Strategies, 2nd edition, Robert J. Wright, Jim Martland, Ann K. Stafford and Garry Stanger, 2006.

Teaching Number sets out in detail nine principles that guide the teaching together with 180 practical, exemplar teaching procedures to advance children to more sophisticated strategies for solving arithmetic problems.

Developing Number Knowledge with 7–11 year olds: Assessment, Teaching and Intervention, Robert J. Wright, David Ellemor-Collins and Pamela Tabor, 2012.

Developing Number Knowledge provides more advanced knowledge and resources for teachers working with older students.

Teaching Number in the Classroom with 4–8 year olds, 2nd edition, Robert J. Wright, Garry Stanger, Ann K. Stafford and Jim Martland, 2014.

This book shows how to extend the work of assessment and intervention with individual and small groups to working with whole classes.

Developing Fractions Knowledge, Amy J. Hackenberg, Anderson Norton and Robert J. Wright, 2016.

Developing Fractions Knowledge provides a detailed progressive approach to assessment and instruction related to students' learning of fractions.

The series provides a comprehensive package on:

1. How to identify, analyse and report students' arithmetic knowledge, skills and strategies.
2. How to design, implement and evaluate a course of intervention.
3. How to include both assessment and teaching in the daily numeracy programme in differing class organizations and contexts.

The series draws on a substantial body of recent theoretical research supported by international, practical application. Because all the assessment and teaching activities have been empirically tested the books are able to show the teacher the possible ranges of students' responses and patterns of their behaviour.

The books are a package for professional development and a comprehensive resource for experienced teachers concerned with intervention and instruction from kindergarten to primary, elementary and secondary levels.

Introduction

Two of the authors of this book have worked extensively on constructivist research into students' development of fractions knowledge. The third author has worked extensively on research, development and implementation related to students' learning of early number and whole number arithmetic. This book applies key ideas and approaches from the latter body of work to the development of a comprehensive guide to developing students' fractions knowledge. As well, the book provides teachers with ways to observe and document students' mathematical activity and thinking, and approaches to instruction that take into account detailed information about students' current levels of knowledge.

The theory and approaches presented in this book have their origins in constructivist teaching experiment research conducted by Leslie Steffe and his collaborators, focusing on students' learning of fractions, as well as the learning of early number and whole number arithmetic.

The Purpose of the Book

The purpose of this book is to provide a detailed and comprehensive guide to classroom and intervention teaching of fractions. This book complements the four earlier books by Bob Wright and colleagues – books that focus on intervention and classroom assessment and teaching of early number and whole number arithmetic:

Early Numeracy: Assessment for Teaching and Intervention

Teaching Number: Advancing Children's Skills and Strategies

Developing Number Knowledge with 7–11 year olds: Assessment, Teaching and Intervention

Teaching Number in the Classroom with 4–8 year olds

The Structure of the Book

Chapter 1 focuses on our general approach to teaching early number and whole number arithmetic and fractions. The chapter begins with three scenarios that portray the ways in which the book can be used by practitioners. Then follow three 'grand organizers' relevant to the approach to instruction that is advocated in the book. First are nine principles of classroom teaching that have been used extensively by the authors and others, as a guide to classroom teaching. Second are the domains of arithmetic knowledge, which show how students' arithmetical knowledge can be organized into progressions of specific topics. Third are the dimensions of mathematizing, that is, bringing advancements in the mathematical sophistication of students' knowledge.

Chapter 2 provides an overview of our general approach to supporting students' development of fractions knowledge. This approach relies upon Steffe's reorganization hypothesis, which describes how students can reorganize their ways of working with whole numbers in order to meaningfully work with fractions. Units coordination plays an important role in both contexts – whole number and fractions. In Chapter 3 we elaborate on units coordination and describe ways that teachers can assess three stages of students' development of units coordination.

Each of Chapters 4 to 12 presents an important topic of developing students' fractions knowledge. There is a common structure for these chapters, consisting of three sections. The first section of each chapter provides an overview of the topic of developing fractions knowledge that is the focus of the chapter. The second section sets out in detail up to six assessment task groups relevant to the topic of the chapter, which can be used to assess, comprehensively, the extent of students' knowledge of the topic. Assessment in this form provides a crucial basis for instruction. An assessment task group is a group of assessment tasks in which all the tasks are very similar to each other. The tasks in each assessment task group focus on a particular aspect of the chapter topic. Each task group includes: a title; the materials used; details of how to present the assessment tasks; and notes on the purpose of the tasks, students' responses and so on. Across Chapters 4 to 12 there is a total of 39 assessment task groups.

The third section of each chapter sets out in detail up to five instructional activities relevant to the topic of the chapter. Each instructional activity has the following six-part format: title; intended learning; instructional mode; materials; description; and notes on responses, variations, and extensions. These activities are designed so that they are easily incorporated into lessons. Across Chapters 4 to 12 there is a total of 39 instructional activities. Many of the assessment task groups can be easily adapted for use as instructional activities, and many of the instructional activities can be easily adapted for use as assessment tasks.

Chapter 13 focuses on relationships between students' fractions knowledge and algebraic reasoning. We give some background on how these relationships have been studied and then discuss the connections between students' fractions knowledge and their work with quantitative unknowns.

The book concludes with a glossary of technical terms used in the text, an appendix that contains templates to be used with some of the assessment tasks and instructional activities, and finally, a list of all of the references cited.

Professional Learning and Three Grand Organizers for Arithmetic Instruction

Chapter 1 outlines ways in which the book can be used to support professional learning focused on strengthening instruction in arithmetic. In order to demonstrate this, three scenarios of professional learning will be included: (i) a mathematics coach working one-on-one to support a teacher in developing an instructional plan for her class; (ii) a school-based mathematics leader working with a team of teachers to revise their school's mathematics programme; (iii) a district-wide Math Recovery® Leader meeting with a team of intervention teachers to develop specialist knowledge aimed at advancing low-attaining students' fractions knowledge. Chapter 1 will also describe links from this text to major generic topics appearing in the current Mathematics Recovery Series, including guiding principles for instruction, domains of arithmetic knowledge and dimensions of mathematizing, such as complexifying, distancing the instructional setting, formalizing, generalizing, notating and unitizing.

Scenario 1: School-Based Mathematics Leader

Mr Phillips is a school-based mathematics leader working with a team of teachers to revise aspects of their school's mathematics programme. The Professional Learning Team (PLT) plans to meet initially for two hours, and then for one hour per week, for a period of six weeks to focus on the teaching of fractions in 4th and 5th grade. Their goal is to develop an instructional programme that (a) is more focused on arithmetical content than their current programme; and (b) enables teachers to take better account of students' current levels of fractions knowledge. Two members of the team have already developed a schedule of assessment tasks drawn from Chapters 3–6, and have administered the schedule individually to each of eight students covering a wide range of attainment levels. At their first meeting the team will review the video recordings of the assessment interviews and develop a simple means of coding students' responses to each of the assessment tasks. Extrapolating from the results of the assessment interviews, they will develop a set of 12 lessons to be taught over a six-week period, based on Chapter 4 (Fragmenting), Chapter 5 (Part–Whole Reasoning), Chapter 6 (Measuring with Unit Fractions) and Chapter 7 (Reversible Reasoning). Instructional plans for each lesson include: (a) a description of the main topic for the lesson; (b) instructional materials; (c) worksheets of relevant exercises incorporating a progression of difficulty; (d) the range of student responses to relevant assessment tasks; and (e) descriptions of common errors, difficulties and misconceptions. For each lesson, the instructional plan includes a 10-minute segment allowing for intensive, targeted intervention with a group of up to six students. This segment focuses on helping students to develop and consolidate their knowledge of the earlier topics of Fragmenting (Chapter 4) and Part–Whole Reasoning (Chapter 5).

Scenario 2: District-Wide Math Recovery® Leader

Ms Gomez is a district-wide Math Recovery® (MR) Leader. In the last two years she has trained two cohorts of 12 MR Intervention Specialists who work in schools across the district. This training has focused on teaching whole number arithmetic across grades K–4 and has involved the following three phases: Phase 1 consists of an initial professional learning programme of three to five days focusing on assessing and profiling students' arithmetical knowledge. Up to 12 students judged as likely to benefit from intensive intervention are individually administered a schedule of assessment tasks. The schedule of tasks has been developed from the sets of assessment tasks in Chapters 3–9. Each teacher's pre-assessment interviews are video-recorded for later analysis. Phase 2 consists of selecting up to four students, each of whom is taught individually for up to five 30-minute sessions per week, for teaching cycles of 12–15 weeks. Phase 2 also includes three or four on-going professional learning sessions during the period of the teaching cycle. Phase 3 involves administering one-on-one post-assessments to the 12 students who underwent the pre-assessment. Each teacher routinely video-records all of their teaching sessions as well as their pre- and post-assessment interviews. In all of the professional learning sessions, the participating teachers present case studies highlighting students' arithmetical strategies and progressions in learning. More detailed descriptions of the professional learning programme described above are available (Wright, 2000, 2003, 2008). This year Ms Gomez will again train two cohorts of Intervention Specialists, but for the first time one cohort will focus on the teaching of fractions at the 5th and 6th grade. In working with this new cohort, Ms Gomez will adopt the year-long professional development model that she has used for several years with other cohorts of teachers in her district.

Scenario 3: Mathematics Coach

Ms Liang is a mathematics coach working one-on-one to support a 5th grade teacher (Ms Koppel). The focus of the coaching is to use video-recorded, one-on-one assessments to advance Ms Koppel's knowledge of fractions pedagogy. Ms Liang meets with Ms Koppel to discuss their working together for a two-week period during which they will develop, administer and review a short schedule of assessment tasks adapted from the assessment tasks in Chapters 5 and 6. These tasks will be designed to assess students' mental actions relating to the fractions topics of partitioning and iterating. Using the assessment schedule, Ms Liang conducts one-on-one, video-recorded assessments with five students from Ms Koppel's class. The five students are selected as representative of a wide range of attainment levels in the learning of fractions. The purposes of the assessments are (a) to inform both coach and teacher of the range and nature of student responses to the assessment tasks and (b) to induct Ms Koppel into the process of using one-on-one assessments to gauge students' current levels of fractions knowledge. In their next meeting they review the video records using the assessment schedule to note students' responses to the tasks. Ms Liang takes the opportunity to highlight the interactive and inquiring nature of the assessment interview. The next day, Ms Koppel conducts one-on-one assessment interviews with a similar group of five students and in their next meeting they use the assessment schedule to review the video records of the second group of students, noting students' responses. As well, Ms Liang reviews Ms Koppel's conducting of the assessment interviews, noting the extent to which Ms Koppel successfully elicits valuable information about the fractions thinking of her students.

Three Grand Organizers for Arithmetic Instruction

In professional learning work with teachers focusing on whole number arithmetic, we developed three grand organizers (see Wright et al., 2006a, 2006b, 2012, 2015). These are: Guiding Principles for Instruction; Domains of Arithmetic Learning; and Dimensions of Mathematizing. In the following sections these grand organizers are described and extended to reasoning and arithmetic involving fractions.

1 Guiding Principles for Instruction

The authors of this book have conducted an extensive range of research and development projects focusing on mathematics pedagogy. Many of these projects involved working in close collaboration with teachers, schools and school systems to advance mathematics instruction. Below we set out nine guiding principles of mathematics instruction which aptly summarize the approach to fractions pedagogy that we advocate. In our collaborative research and development work, these principles have been applied extensively to guide the teaching of mathematics.

[1]   The teaching approach is inquiry-based, that is, problem-based. Students routinely are engaged in thinking hard to solve fractions problems that for them are quite challenging.

[2]   Teaching is informed by an initial, comprehensive assessment and on-going assessment through teaching. The latter refers to the teacher's informed understanding of students' current knowledge and problem-solving strategies, and continual revision of this understanding.

[3] Teaching is focused just beyond the 'cutting-edge' of students' current knowledge.

[4] Teachers exercise their professional judgement in selecting from a range of teaching procedures each of which involves particular instructional settings and tasks, and varying this selection on the basis of on-going observations.

[5] The teacher understands students' fractions strategies and deliberately engenders the development of more sophisticated strategies.

[6] Teaching involves intensive, on-going observation by the teacher and continual micro-adjusting or fine-tuning of teaching on the basis of that observation.

[7] Teaching supports and builds on students' intuitive strategies and these are used as a basis for the development of written forms of mathematics that accord with students' verbally-based and pictorially-based strategies.

[8] The teacher provides students with sufficient time to solve a given problem. Consequently students are frequently engaged in episodes that involve sustained thinking, reflection on their thinking and reflection on the results of that thinking.

[9] Students gain intrinsic satisfaction from their problem-solving, from their realization that they are making progress and from the verification methods they develop.

Each of these principles is now discussed in more detail.

Principle 1

The teaching approach is inquiry-based, that is, problem-based. Students routinely are engaged in thinking hard to solve fractions problems that for them are quite challenging.

The inquiry-based approach to teaching mathematics is sometimes referred to as learning through problem-solving or problem-based learning. In this approach, the central learning activity for students is to solve tasks that constitute genuine problems – problems for which the students do not have a ready-made solution. What follows is that the issue of whether a particular task is appropriate as a genuine problem largely depends on the extent of their current knowledge.

Principle 2

Teaching is informed by an initial, comprehensive assessment and on-going assessment through teaching. The latter refers to the teacher's informed understanding of students' current knowledge and problem-solving strategies, and continual revision of this understanding.

Assessment for providing specific and detailed information to inform instruction is the critical ingredient in our approach to teaching mathematics. Through assessment, teachers can develop a working model of students' current knowledge of fractions. Thus it is essential to conduct a detailed assessment of their current knowledge, and to use the results of that assessment in designing instruction. As well, it is essential to revise and update one's understanding of students' knowledge through on-going, close observation of their responses to assessment tasks. The chapters that follow contain detailed descriptions of assessment tasks and notes on their use. These have the explicit purpose of informing the design of instruction. The second aspect of this principle, on-going assessment through observation and reflection, is equally as important as initial assessment and a teacher's understanding of a student's current knowledge can always be deepened.

Principle 3

Teaching is focused just beyond the 'cutting-edge' of students' current knowledge.

This principle accords with Vygotsky's (1978: 84–91) notion of the zone of proximal development and Steffe's (1991) notion of the zone of potential construction; that is, instruction should be focused just beyond the student's current levels of knowledge in areas where that student is likely to learn successfully through sound teaching. This principle is very important in our focus on fractions instruction, and it highlights the importance of assessment to inform teaching. As well, this principle highlights the importance of teachers' evolving understanding of students, and of the idea that productive struggle is essential in learning. Assessment provides the teacher with a profile of students' knowledge and the teacher focuses instruction to provide opportunities for students to move beyond their current levels of knowledge.

Principle 4

Teachers exercise their professional judgement in selecting from a range of teaching procedures each of which involves particular instructional settings and tasks, and varying this selection on the basis of on-going observations.

This principle highlights the need to develop a range of instructional procedures and to understand the role of each procedure, in terms of its potential to bring about advancements in students' current knowledge. The chapters that follow contain detailed descriptions of instructional activities that can be used to develop appropriate teaching procedures. Also, the assessment tasks provided in each chapter constitute an additional source of instructional procedures because the tasks are easily adapted for instruction.

Principle 5

The teacher understands students' fractions strategies and deliberately engenders the development of more sophisticated strategies.

This principle highlights the need for teachers to have a working model of students' knowledge of fractions and the ways in which their knowledge typically progresses. Each of the chapters that follow provides a detailed overview of the development of aspects of fractions knowledge. Our belief is that, through reading, observing and reflecting, in conjunction with their teaching practice, teachers can significantly develop their knowledge of fractions pedagogy.

Principle 6

Teaching involves intensive, on-going observation by the teacher and continual micro-adjusting or fine-tuning of teaching on the basis of that observation.

This principle highlights the importance of observational assessment in determining students' specific learning needs, and the need for this assessment to be on-going and to lead to action, that is, the fine-tuning of instruction on the basis of on-going assessment.

Principle 7

> Teaching supports and builds on students' intuitive strategies and these are used as a basis for the development of written forms of mathematics that accord with students' verbally-based and pictorially-based strategies.

This principle highlights that students express and develop their fractions knowledge, at least initially, via verbal expressions and images. That is, they develop their knowledge via images if they are supported to work with and develop their ideas with pictures and manipulatives that open possibilities for them to visualize quantities and verbalize their findings.

Principle 8

> The teacher provides students with sufficient time to solve a given problem. Consequently students are frequently engaged in episodes that involve sustained thinking, reflection on their thinking and reflection on the results of that thinking.

In our research and development work in mathematics pedagogy, we emphasize the importance of sustained thinking and reflection for the learning of mathematics. The topic of fractions is well suited to significant problem-solving by students. This problem-solving and the mental processes of thinking hard and reflecting during problem-solving are, we believe, a fundamental aspect of learning fractions.

Principle 9

> Students gain intrinsic satisfaction from their problem-solving, from their realization that they are making progress and from the verification methods they develop.

This principle relates to Principle 8. Our experience in working closely with teachers and students for many years on the topics of whole number arithmetic and fractions is that when young students work hard at problem-solving and that problem-solving is successful, this is typically a very positive experience for the learner. To go further, we argue that this kind of learning constitutes a kind of cognitive therapy, having intrinsic rewards beyond teacher affirmation and peer recognition.

2 Domains of Arithmetic Knowledge

In our research and development work related to mathematics pedagogy, we have found it helpful to organize instruction in whole number arithmetic into a framework of bands and domains. In this book we extend this framework to include fractions pedagogy (see Table 1.1).

Table 1.1   Knowledge of whole number arithmetic

Band 1 – Very early number
Band 2 – Early number
   Number words and numerals
   Early counting, addition and subtraction
   Structuring numbers 1 to 10
Band 3 – Middle number
   Advanced number words and numerals
   Structuring numbers 1 to 20
   Conceptual place value
   Addition and subtraction to 100
   Multiplicative strategies
   Multiplication and division basic facts
Knowledge of fractions
Band 4 – Fractions as solely part–whole concepts
Band 5 – Fractions as early measures
Band 6 – Fractions as fractional numbers

This framework of domains of knowledge provides a basis for extension from whole number arithmetic to the arithmetic of fractions. Each of these bands and domains is briefly described below. More detailed descriptions of these domains are available (e.g. Wright et al., 2012, 2015).

Band 1 – Very Early Number

In the first three or four years of life children develop nascent number knowledge. This earliest domain of number knowledge can be organized into subdomains. First, children hear number words and begin to organize them as a special set of words that often occur in a sequence, first forwards and then backwards as well. Second, children begin to learn to name or read numerals – initially this mainly involves one-digit numerals (e.g. 5, 8, 3) and is gradually extended beyond 10. Third, children acquire early notions of counting – to count is to say the words from one to four, for example while making pointing actions, and emphasizing the last number word, but the points might not necessarily be in one-to-one coordination with the number words. Fourth, children also begin to ascribe number to the simple configurations that occur on dice, dominoes and so on. All of this emerging knowledge provides an excellent basis for developing more advanced knowledge.

Band 2 – Early Number

We use the term 'early number' to label the arithmetic knowledge that children typically develop in the first two or three years of school (Table 1.1). We organize this knowledge into three domains and each of these can be linked to the subdomains of very early number (see above). These domains are (a) number words and numerals; (b) early counting, addition and subtraction; and (c) **structuring numbers** 1 to 10. Our co-series book entitled *Teaching Number* *in the Classroom with 4–8 year olds* (2nd edition, 2014) addresses these domains in great detail, and so only brief descriptions are provided here (see also Wright, 2013). The domain of number words and numerals includes children's learning of forward number word sequences, the number word after a given number word, backward number word sequences, the number word before a given number word, and learning to read and write numerals. The domain of early counting, addition and subtraction includes children's use of counting, typically by ones, in increasingly sophisticated ways to solve tasks involving how many items in a collection, how many in all in the case of two collections, how many remaining and so on. In this domain children solve counting-based tasks typically in the ranges 1 to 10 and 1 to 20. Important to realize is the clear distinction between the two domains just described (Wright, 2013). Finally, the domain of **structuring numbers** 1 to 10 involves developing early arithmetical knowledge that does not involve counting-by-ones. This includes topics such as: (a) partitioning 5, partitioning 10, the doubles of numbers in the range 1 to 5; and (b) using that knowledge to solve addition and subtraction tasks in the range 1 to 10.

Band 3 – Middle Number

We use the term 'middle number' to label the whole number arithmetic knowledge that children typically develop in the range 2nd to 5th grade (Table 1.1). We organize this knowledge into six domains and these can be linked to the domains of early number (Table 1.1). These domains are (a) Advanced Number Words and Numerals; (b) **Structuring Numbers** 1 to 20; (c) Conceptual Place Value; (d) Addition and Subtraction to 100; (e) Multiplicative Strategies; and (f) Multiplication and Division Basic Facts. Our co-series book entitled *Developing Number Knowledge: Assessment, Teaching and Intervention with 7–11 year olds* (2012) addresses these domains in great detail, and so only brief descriptions are provided here. The domains of advanced number words and numerals and **Structuring Numbers** 1 to 20 extend the corresponding domains in the Early Number Band. The domain of Conceptual Place Value encompasses a novel approach to place value instruction. This domain focuses on developing increasingly sophisticated strategies for incrementing and decrementing by units of 1s, 10s, 100s and so on. Sound knowledge of this domain provides an important basis for the domain of Addition and Subtraction to 100, that is, developing advanced strategies for mental addition and subtraction, such as the jump and split kinds of strategies (Wright et al., 2012). The last mentioned domain also includes learning to add and subtract to and from a decuple without counting by ones. The domains of Multiplicative Strategies and Multiplication and Division Basic Facts include: (a) learning to structure numbers multiplicatively, for example 24 is 4×6, 12×2 and so on; (b) developing increasingly sophisticated strategies for multiplication in the range 1–100 and beyond; and (c) meaningful habituation of the basic facts of multiplication and division (Wright et al., 2012). Collectively, the domains of Middle Number provide the basis for extending whole number arithmetic knowledge to fractions knowledge, and the latter is the focus of subsequent chapters of this book.

Band 4 – Fractions as Solely Part–Whole Concepts

Band 4 encompasses students who have constructed Parts Within Wholes and Parts Out of Wholes Fraction Schemes where fractions are determined based on the number of equal parts in the fraction and the whole. Band 4 is the focus of Chapters 4 and 5.

Band 5 – Fractions as Early Measures

Band 5 encompasses students who are beginning to understand fractions as sizes relative to the whole. These students are moving away from solely part–whole concepts, but they have not yet fully developed fractions as measures or numbers. Band 5 is the focus of Chapters 6 and 7, and it is also addressed in Chapters 9–12.

Band 6 – Fractions as Numbers

Band 6 encompasses students who have constructed Iterative Fraction Schemes and, thus, conceive of fractions truly as numbers and as extensive quantities (i.e. measures). Band 6 is the focus of Chapter 8 and is also addressed in Chapters 9–12.

3 Dimensions of Mathematizing

Progressive mathematization refers to progression to new learning that is mathematically more sophisticated than current learning (Gravemeijer et al., 2000). In our work focusing on developing whole number arithmetic knowledge, we developed a model of 11 dimensions of mathematizing:

1. Complexifying
2. Decimalizing numbers
3. Distancing the instructional setting
4. Extending the range of numbers
5. Formalizing
6. Generalizing
7. Grounded habituation
8. Notating
9. Refining computational strategies
10. Structuring numbers
11. Unitizing

Our co-series book entitled *Developing Number Knowledge: Assessment, Teaching and Intervention with 7–11 year olds* (2012) addresses these dimensions in detail (see also Ellemor-Collins and Wright, 2011). In this book we describe these dimensions from the perspectives of whole number arithmetic and students' developing fractions knowledge. As well, examples of mathematizing are highlighted in the instructional activities.

Complexifying

Complexifying refers to making an instructional situation more complex arithmetically. Examples include the following: progressing from comparing unit fractions to comparing non-unit fractions; progressing from adding or multiplying proper fractions to adding or multiplying improper fractions; progressing from comparing improper fractions just larger than one to improper fractions much larger than multiple wholes; and progressing from dividing whole numbers by fractions to dividing mixed numbers by fractions.

Decimalizing

Decimalizing refers to coming to know the base-ten structure of the numeration system. Thus mathematizing by **decimalizing** is a particular form of mathematizing by structuring (see below). Initially, students begin to realize the additive structure of two-digit numbers – a decuple plus a number in the range 0–9 (e.g. 46 consists of 40 and 6). This initial **decimalizing** of numbers is extended to a structure that is both additive and multiplicative (e.g. 346 consists of $3 \times 100 + 4 \times 10 + 6 \times 1$). At much the same time, this knowledge is extended to three- and four-digit numbers and beyond. **Decimalizing** also involves learning the system of decimal fractions and the links between decimal fractions and common fractions. This can lead to the realizations that every common fraction can be expressed as a terminating or repeating decimal and vice versa. Finally, arithmetic involving common fractions provides the important basis for progressing from arithmetic to algebra (see Chapter 13).

Distancing the Instructional Setting

We use the term '*instructional setting*' to describe the materials that teachers use to support students' reasoning about fractions. In some instructional situations it can be helpful to screen the materials and perhaps also to screen the notations students generate. The process of screening can support students' visualization associated with reasoning about fractions, which in turn can support their development of fractions knowledge. For example, a teacher asks students to visualize taking a unit fraction of a unit fraction after having made drawings.

Extending the Range of Numbers

When presenting arithmetical tasks to students, there is always scope to move beyond an initial range of numbers. In the case of addition with whole numbers for example, students' learning focuses initially on adding two numbers in the range 1–10, and this can extend to 20, to 100 and so on. In the case of fractions instruction, the initial step from addition involving whole numbers to addition involving simple fractions is an example of **extending the range of numbers**. Within fractions, instruction can extend from unit fractions to non-unit proper fractions and further to improper fractions.

Formalizing

Formalizing refers to students' progression from initial ways of reasoning and solving problems involving fractions to forms of reasoning typical of expert adults. This could involve students' progressing from intuitive procedures to formal algorithms when operating on fractions. For example, in adding two fractions, say $\frac{2}{3}$ and $\frac{3}{5}$, students might need to rely on drawings, at first, in order to find a common measurement unit ($\frac{1}{15}$) – partitioning each of the thirds into five equal parts and partitioning each of the fifths into three equal parts. **Formalizing** would involve representing that activity in a number sentence: $\frac{2}{3} + \frac{3}{5} = \frac{10}{15} + \frac{9}{15} = \frac{19}{15}$.

Generalizing

Generalizing refers to progressing from arithmetical to algebraic reasoning about fractions. Examples include **generalizing** in situations that involve comparing unit fractions according to the size of the denominator; describing a process for generating fractions equivalent to a fraction expressed in low-est terms; and describing a process for multiplication or division involving two fractions.

Grounded Habituation

A student who, when asked '8 plus 9', almost immediately answers '17', and similarly answers other additions in the range 1 to 20, has *habituated* addition facts. When asked to justify the answer the student might say, 'I know 8 + 8 makes 16 and I need to add one more'. Thus a student who has **grounded habituation** not only knows the basic addition facts but can also devise explanations of answers that typically draw on additive structure such as five-structure, ten-structure and doubles structure of even numbers. In the case of fractions, **grounded habituation** might involve habituated knowledge such as $\frac{1}{2} + \frac{1}{4} = \frac{3}{4}$, along with being able to explain that $\frac{3}{4}$ is the answer because $\frac{1}{2}$ and $\frac{2}{4}$ are commensurate fractions.

Notating

In some instructional situations involving fractions, teachers ask their students to write or draw in order to symbolize their reasoning or to show a problem-solving strategy. This can lead to students developing an informal notation that they incorporate into their learning about frac-tions. We refer to this student activity as **notating**. **Notating** in this way can support advance-ments in students' conceptual knowledge of fractions. For example, a student might draw an arrow between a rectangle representing $\frac{1}{5}$ and a rectangle representing the whole, and then might label the rectangle '5 times' to notate the size relationship between $\frac{1}{5}$ and the whole. **Notating** that both students and teachers do should be a trace of their reasoning with pictures and mental images (representations of quantities), and then eventually come to stand in for reasoning.

Refining Computational Strategies

In the case of whole number arithmetic, students can learn intuitive mental strategies for adding and subtracting in the range 0–100 which are relatively sophisticated. For example, a student solves 48 + 35 in the following three steps: 48 + 30 = 78; 78 + 2 = 80; 80 + 3 = 83 (referred to as a 'jump' strategy). At an earlier time the student might have added three 10s singly (48 + 10; 58 + 10; 68 + 10). The former strategy involves curtailment of the latter strategy. This exemplifies *refining a computational strategy*. In the case of arithmetic involving fractions there are many examples of refinement of a computational strategy. For example, solving $\frac{7}{8} + \frac{5}{6}$ might involve an elaborated procedure to determine that twenty-fourths (or forty-eighths) can serve as a com-mon measurement unit for eighths and sixths. Mathematizing in this case could involve finding the common measurement unit in one step. More broadly, learning the standard algorithms for adding, subtracting, multiplying and dividing involving fractions can involve the progressive refinement of computational strategies.

Structuring Numbers

Structuring occurs when the learner mentally imposes some form of arithmetical structure on numbers. A young learner who can say the number words from 1 to beyond 12, and can count a small collection of objects, might regard 17 as consisting of 17 ones. Contrast this with the learner who regards 17 as 10 + 7, 10 + 5 + 2, 8 + 8 + 1, or 20 − 2 − 1. The latter learner has what is called additive structure. Similarly, the student who can regard 24 as 2 × 12, 4 × 6, 240 ÷ 10 and so on, has developed multiplicative structure. Structuring in the case of whole numbers can be extended to additive and multiplicative structuring involving fractions; for example, regarding seven-fifths as both $\frac{7}{5}$ and $1\frac{2}{5}$.

Unitizing

Unitizing refers to the mental act of regarding a collection as a unit. The collection could be a strip of ten dots, or three sevenths of a piece of string. The dimension of **unitizing** and units coordination are referred to extensively in this book, as explained in Chapters 2 and 3.

2

From Whole Numbers to Fractions

For many students during their early elementary school years fractions present a notorious challenge and perpetual struggle. Students often try to memorize the rules for adding, comparing, multiplying and dividing fractions, but none of it makes much sense. Even students who successfully memorize rules do not necessarily develop fractions knowledge that is flexible and robust. Our approach to addressing the dilemma builds on students' whole number knowledge and, specifically, the ways that students build and work with units.

This chapter begins with an overview of Steffe and Olive's (2010) reorganization hypothesis (also see Steffe, 2002), which describes how students can reorganize their whole number knowledge to generate fractions. We will elaborate on five mental actions that support students' ways of working with units across whole number and fractions contexts: **unitizing**, fragmenting/partitioning, iterating, disembedding and distributing. Then we describe the role of these mental actions in coordinating units. Finally, we outline the fractions content chapters (Chapter 4–12), which are organized around students' stages of units coordination. In Chapter 3 we provide tools for assessing students' stages of units coordination.

The Reorganization Hypothesis

Why do students continually struggle with fractions? One possible explanation is that the rules for working with fractions conflict with the rules for working with whole numbers. For example, 7 is bigger than 6 and yet $\frac{1}{7}$ is smaller than $\frac{1}{6}$. This fact is counterintuitive for students who do not yet understand the reciprocal relationship between the number of parts in a whole and the size of each part: namely, the more equal parts there are in a whole, the smaller the parts will be. Also, with the exception of 0 and 1, multiplying two whole numbers always results in a number greater than the two numbers being multiplied, but this is not always the case with fractions (e.g. $\frac{2}{3}$ times $\frac{1}{2}$ is $\frac{1}{3}$, which is smaller than the two other numbers). When multiplying fractions, students can simply memorize the rule of multiplying the numerators and denominators as if multiplying whole numbers 'twice,' but when adding fractions there are new rules. Examples like these have been used to support the interference hypothesis, which states that students' whole number knowledge often interferes with their fractions knowledge (Streefland, 1991). However, when we focus on students' mental actions, we find much more continuity in their development of fractions knowledge from their whole number knowledge.

The reorganization hypothesis states that the ways students work with whole numbers influence the ways they learn to work with fractions (Steffe, 2002). In fact, students perform many of the same mental actions to make sense of fractions that they perform to make sense of whole numbers. Here, we describe five of these mental actions and their connection to both whole numbers and fractions.

Unitizing

Consider the two adjectives in the phrase 'six red roses'. Both 'six' and 'red' describe the roses, but not in the same way. We cannot perceive 6 the way we perceive red: we have to make 6 through mental action, beyond the psychological processes involved in perception. The same can be said for any number, including 1. Anything can be 1. Even the six red roses can be 1 – one half dozen – if we make it that way in our minds. Actually, we might treat the six red roses as a unit of 6 and a unit of 1 at the same time, making it into a composite unit – a unit composed of other units. The mental action we perform in making these units is called **unitizing**.

Unitizing arises through students' actions of counting and working with whole numbers, as described in *Developing Number Knowledge* (Wright et al., 2012). In particular, students might begin treating 6 as a unit by segmenting their counting sequence into collections of six 1s and then treating these segments as something to count ({1, 2, 3, 4, 5, 6}; {7, 8, 9, 10, 11, 12}; ...). These collections become composite units when students can count them without losing track of the units of 1 that comprise them, which makes it possible for students to double count. For instance, students who build composite units of 6 can determine the value of three 6s by double-counting as follows: '1, 2, 3, 4, 5, 6 is 1; 7, 8, 9, 10, 11, 12 is 2; and 13, 14, 15, 16, 17, 18 is 3. So three 6s is 18.'

With fractions, students need to work with different kinds of units, but the mental action of **unitizing** is the same. The whole is a particularly important unit. Essentially a unit of 1 that can be counted, the whole is also a composite unit comprised of smaller units, including unit fractions. For example, students can unitize the whole and repeat it five times to make five wholes, but they might also consider the whole as a composite unit made up of five $\frac{1}{5}$ units (see Figure 2.1). In this sense, we can think of unit fractions, like $\frac{1}{5}$, as units of measure. If a student has unitized $\frac{1}{5}$ she can measure off other lengths or other quantities with that fractional unit, much like she would with a centimetre, a litre, or any other unit of measure.

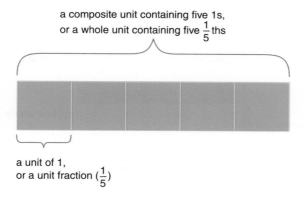

a composite unit containing five 1s,
or a whole unit containing five $\frac{1}{5}$ths

a unit of 1,
or a unit fraction ($\frac{1}{5}$)

Figure 2.1   Whole number and fractional units

Fragmenting and Partitioning

In Chapter 4 we describe partitioning as a special case of fragmenting – breaking a whole into parts. Partitioning occurs when students attend to making the parts equal while exhausting the whole (i.e. there is no leftover piece). Although we often think about this mental action in the context of fractions, students begin partitioning collections of objects with whole numbers. In particular, students can develop the mental action of partitioning through experiences of fair sharing, or *equal sharing*. Equal sharing can occur in the context of sharing a collection of objects among a fixed number of people so that every person receives the same number of objects. For example, in order to share 12 coins fairly among four friends, the friends might each take one coin at a time until none are left; after three turns, each friend will have three coins with none left over. However, partitioning requires a bit more: students need to see the four units of 3 within the 12. Thus, **unitizing** plays a role in partitioning.

In the context of fractions, the whole could be a continuous unit (such as a granola bar) rather than a collection of discrete units (such as 12 coins). Partitioning a continuous unit results in equally sized unit fractions rather than equal groups. Students can still attempt equal sharing by breaking off small pieces and distributing them until the whole is exhausted, but again, partitioning requires more. Students need to see the fractional units within the whole, just as they need to see the composite units when working with discrete collections like 12. Students can begin to see these fractional units in either of two ways: simultaneous partitioning or equi-partitioning.

Simultaneous partitioning involves projecting a composite unit into the continuous whole. For example, if a student wants to share a granola bar among four friends and she has already developed a composite unit of 4, she might imagine the composite unit being spread out evenly within the whole bar. Figure 2.2 illustrates the idea by showing four equally spaced dots being projected into the whole so that we can imagine how to partition the whole into four equal parts.

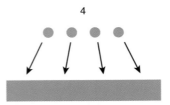

Figure 2.2 Simultaneous partitioning

Equi-partitioning would involve creating a single part, as an estimated equal share, and then iterating that part four times to see whether it exhausts the whole. The next section elaborates on the mental action of iterating and then demonstrates its use in equi-partitioning. Students' development of the mental action of partitioning is the focus of Chapter 4.

Iterating

Consider the task shown in Figure 2.3. A student might solve the task by making connected copies of the small bar until reaching the length of the long bar. When students can carry out this action in imagination, the mental action is referred to as iterating. Like partitioning, the result of iterating involves two levels of units: the unit being iterated (in this case, the length of the small bar) and the composite unit comprised of those iterations (in this case, the length of the long bar).

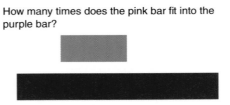

Figure 2.3 Iterating task

Iterating also occurs in whole number contexts. Students might iterate a unit of 1 five times in order to produce a composite unit of 5. In fact, 5 would be considered a composite unit precisely because it is composed of those five units of 1 that are created through iteration. The mental action of iterating produces a 1-to-5 relationship between the quantity being iterated (1) and the quantity formed through iteration (5). We can see the origins of multiplicative reasoning in this kind of action: in addition to being a collection of five 1s, 5 is also five times as big as any of those 1s.

Partitioning and iterating have a special relationship in that students can begin to under-stand these two mental actions as inverses of one another. For example, consider the situation of equi-partitioning again, in which a student wants to share a granola bar among four friends. The student might produce an initial estimate and then iterate that piece four times. If these four iterations exhaust the whole, as illustrated in Figure 2.4, the student might understand that they have produced an equal share. As teachers, we might take the implication for granted, but as we will discuss in Chapter 6, many students have yet to coordinate partitioned units and iterated units in this way. In general, they do not understand that if a whole is partitioned into *n* equal pieces, they can iterate any one of those pieces *n* times to get back to the whole.

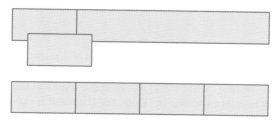

Figure 2.4   Equi-partitioning

Disembedding

We have seen that partitioning and iterating are inherently connected to the idea of composite units and coordinating those units with their constituent units (e.g. units of 1). Disembedding is another mental action that involves this kind of coordination. Specifically, it involves taking a part out of a whole without destroying the whole. Once again, students can develop this mental action in the context of whole numbers.

Consider the numbers 5 and 7. We know that 7 is bigger than 5 because 7 comes after 5 in the counting sequence. For example, when asked which number is bigger, a kindergarten student named Ella responded that 7 was bigger, 'because 1, 2, 3, 4, 5, and you haven't heard 7 yet'. But 7 does not merely follow 5; 7 contains 5 in that 5 is a part of 7. These part–whole relations have obvious implications for fractions. As described in Chapter 5, students' initial conceptions of fractions are generally part–whole conceptions; they generally consider $\frac{5}{7}$ as five parts within or out of seven equal parts in the whole. However, without disembedding, their part–whole conceptions might seem idiosyncratic to the teacher. For example, the students might agree that Figure 2.5 represents $\frac{5}{7}$, but how they consider that representation can diverge depending on whether they disembed the five parts from the seven parts in the whole.

Figure 2.5   Shading $\frac{5}{7}$

When asked to identify $\frac{5}{7}$ in the figure, some students will respond that the entire picture is $\frac{5}{7}$ (Olive and Vomvoridi, 2006). Other students understand that $\frac{5}{7}$ only refers to the shaded part. These students might further understand that any quantity of that size, even when removed from the seven-part whole, is $\frac{5}{7}$ of that whole. This conception relies on disembedding the part ($\frac{5}{7}$) from the whole so that the relationship between the part and the whole is not lost just because the part is no longer drawn within the whole. Supporting and leveraging students' mental actions of iterating and disembedding is the focus of Chapter 6.

Distributing

Students who produce composite units as iterations of a unit of 1 might begin to coordinate multiple composite units by distributing the 1s contained within one composite unit across the 1s contained in the other composite unit (Steffe, 1992). For example, a student might distribute three units of 1 across each of the 1s in 8, as illustrated in Figure 2.6. The result is eight 3s, or twenty-four 1s.

Figure 2.6   Distributing three units of 1 across eight units of 1

The connection to whole number multiplication is obvious, but units distributing also plays a role in multiplying fractions. Consider the same distribution of units in a continuous (rather than discrete) context, where each of the eight parts is partitioned into three parts (see Figure 2.7).

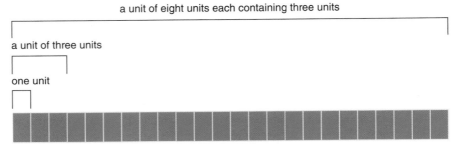

Figure 2.7   Distributing partitions of three across a partition of eight

Source: Norton, Boyce, Phillips et al., 2015

As mentioned previously, multiplying fractions can seem quite different from multiplying whole numbers. Students can think of whole number multiplication as repeated addition or forming groups of groups (e.g., 8 times 3 is eight groups of 3, as shown in Figures 2.6 and 2.7).

However, this heuristic does not work very well for fractions. Rather, students can think of fraction multiplication as taking a fraction of another fraction – composing two fractions. For example, consider the product of $\frac{1}{7}$ and $\frac{1}{5}$, illustrated on the left side of Figure 2.8. Students can think of this product as $\frac{1}{7}$ of $\frac{1}{5}$ of a whole: take $\frac{1}{5}$ of the whole, then take $\frac{1}{7}$ of the result (one of the small parts at the bottom left of Figure 2.8), and then compare that to the whole. Likewise, $\frac{1}{7}$ of $\frac{2}{5}$ would be $\frac{1}{7}$ from each of two $\frac{1}{5}$ parts (see the right side of Figure 2.8).

Figure 2.8 $\frac{1}{7}$ of $\frac{1}{5}$ of a whole *(left)* and $\frac{1}{7}$ of $\frac{2}{5}$ of that whole *(right)*

Both cases involve taking a fraction of a fraction of a whole. In this sense, fraction multiplication is the composition of two fractions – a fraction of a fraction. The composition involves distributing the fractional units from one fraction across the fractional units of another fraction (Hackenberg and Tillema, 2009). We discuss fraction multiplication further in Chapter 10, and we address addition, subtraction and division in Chapters 11 and 12.

Table 2.1 summarizes the mental actions discussed here. In the next section we focus on their connection to units coordination.

Table 2.1 Mental actions for making sense of whole numbers and fractions

Mental action	Description	Example
Unitizing	Treating a quantity as a single unit	Taking a collection of 12 chips as a whole
Partitioning	Producing equally sized parts or groups	Sharing a bar equally among four friends
Iterating	Producing connected copies of a quantity	Repeating a $\frac{1}{5}$ part five times to make the whole
Disembedding	Taking out a part of a whole without destroying the whole	Taking $\frac{5}{7}$ out of $\frac{7}{7}$ while maintaining $\frac{5}{7}$ as part of that whole
Distributing	Inserting units from one composite unit into units of another composite unit	Partitioning each $\frac{1}{5}$ part of the whole into seven parts to make parts that are $\frac{1}{35}$ of the whole

Units Coordination

Units coordination refers to the ways students build and work with units (Steffe, 1992). We have already seen some examples of this in the context of fractions:

1) coordinating the part taken out of the whole with the whole from which it was taken (e.g. viewing $\frac{3}{5}$ as three equal parts out of five equal parts in the whole);
2) coordinating a unit fraction with the whole through its multiplicative relation (e.g. viewing the whole as a unit that is five times as big as the fractional unit, $\frac{1}{5}$);
3) coordinating unit fractions distributed across other units fractions (e.g. determining the value of $\frac{1}{7}$ of $\frac{1}{5}$ by distributing $\frac{7}{7}$ within each $\frac{1}{5}$ in the whole).

The first example involves disembedding; the second example involves iterating; and the third example involves distributing. In this section, we focus on units coordination in whole number contexts, where these mental actions first arise. Chapters 4–12 will elaborate on units coordination in various fractions contexts.

Steffe's (2002) reorganization hypothesis stipulates that the ways students build and work with units in whole number contexts influence the ways they coordinate units in fractions contexts. So, in preparing to teach students fractions, it is helpful to understand how students coordinate units in whole number contexts. We can identify three stages in students' units coordination abilities, as distinguished by the number of levels students coordinate: units of 1, units of units (e.g. composite units) and units of units of units (e.g. composite units distributed across other composite units).

Levels of units refer to the embeddedness of units within other units. For example, when students work with composite units, they are working with two levels of units because there are other units embedded within the composite unit. When students do something with the composite unit, they are simultaneously doing something with those constituent units. If students have to carry out mental actions separately with each level of units, they are not really working on the composite unit.

Stages of units coordination refers to students' development in working with one, two, or three levels of units at the same time. Students generally progress from one stage to the next through a sequence of mental actions that they carry out. Once they have organized those mental actions so that they no longer need to carry them out in order to maintain two or three levels of units, we can say that they have progressed to the second or third stage, respectively.

Consider how students might respond to the task shown below, with reference to Figure 2.9. Note that the bars are not drawn to scale, which is intended to discourage students from visually estimating the unknown relation.

'If the red bar fits into the yellow bar three times, and the yellow bar fits into the blue bar four times, how many times would the red bar fit into the blue bar?'

At Stage 1, students can work flexibly with units of 1 and may even work with composite units as collections of 1s. However, there is no multiplicative relationship working across these two levels of units. In other words, they have not unitized – and therefore do not conceptualize – a

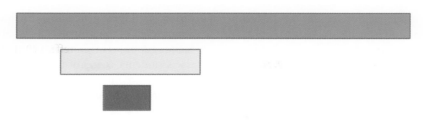

Figure 2.9   Bars task

Source: Norton, Boyce, Phillips et al., 2015

composite unit, say 5, as a unit that is 5 times as big as 1. In the bars task, these students are unlikely to use the given relations. Instead, they might take the red bar as a unit of 1 and iterate its length across the length of the blue bar to determine the unknown relation. Thus they will often rely on the figurative material provided in the picture.

At Stage 2, students can unitize the result of iterating another unit and, thus, they can take two-level multiplicative relations as given. They can meaningfully use the given relations (e.g. three red bars fit into the yellow bar) in further activity. However, they have to carry out some kind of activity to coordinate the two given relations and build a third level of units. For example, they might treat the yellow bar as a unit of three 1s (red bars) and mark off that composite unit four times (as specified by the given relation between the yellow bar and the blue bar) in order to determine that the red bar fits into the blue bar 4 times 3 times. While carrying out the activity of marking off the composite unit, they are partitioning the blue bar into four parts and distributing three units of the red bar within each partition.

At Stage 3, students can put the given relations together into a unit of units of units: the blue bar as four yellow bars, each of which is three red bars. These students can immediately apprehend the unknown relationship as a multiplicative coordination of the given two relationships without the need to perform iterating or distributing actions, physically or mentally. Table 2.2 summarizes the three stages of units coordination.

Table 2.2   Stages of units coordination

	Students' unit structures	**Students' reasoning on the bars task**
Stage 1	Students can take one level of units as given, and may coordinate two levels of units in activity	Students mentally iterate the red bar, imagining how many times it would fit into the blue bar. This action might be indicated by head nods or sub-vocal counting.
Stage 2	Students can take two levels of units as given, and may coordinate three levels of units in activity	Students mentally iterate the yellow bar four times, with each iteration representing a 3. This action might be indicated by the student uttering '3, 3, 3, and 3; 12'.
Stage 3	Students can take three levels of units as given, and can thus flexibly switch between three-level structures	Students immediately understand that there are four 3s in the blue bar. This assimilation of the task might be indicated by an immediate response of '12', buttressed by an argument that 12 is four 3s.

To summarize this section, mental actions of **unitizing**, partitioning, iterating, disembedding and distributing enable students to coordinate the various levels of units involved in whole numbers and fractions. Units coordination is central to Steffe's (2002) reorganization hypothesis. In Chapter 3 we introduce a method for teachers to assess students' stages of units coordination. Subsequent chapters describe ways that teachers can use students' units coordinating actions and the associated mental actions to support students' development of specific fractions ideas.

Introduction to Content Chapters

Assessments of students' stages of units coordination establish a launching point for instruction, whole class or individual. For whole-class instruction, teachers might consider homogeneous grouping so that students at the same stage can work together on tasks associated with that stage. On the other hand, teachers could consider heterogeneous grouping so that students at higher stages can challenge students working at lower stages while developing their abilities to articulate their reasoning.

Whether working with individuals, groups, or a whole class at once, teachers should purposefully select tasks that will appropriately challenge students. Appropriate tasks are those that the students understand but without a ready-made response; these tasks require a student to perform some kind of action in order to figure them out. This action becomes the basis for new concepts. If the tasks are not demanding enough, students may practise existing concepts but will not learn new ones. If the tasks are too demanding, students may have no way to work on them – they will not know what mental actions to perform to solve them.

This book is organized into chapters that should help teachers select appropriately challenging tasks for their students. Assessments of students' stages of units coordination provide a starting point in Chapter 3. Chapters 4–12 are organized by the three stages. Tasks are further guided by assessments of students' *schemes*, i.e. their ways of working with mathematical objects, such as whole numbers and fractions. For the benefit of teachers in the United States, many of the instructional tasks are explicitly connected to Common Core State Standards (2010).

Stage 1

Chapters 4 and 5 address concepts that students working at Stage 1 might develop. Although students at Stage 1 will be limited in their abilities to work with fractions, there are activities in which teachers can engage them to (1) begin developing fractions knowledge and (2) promote further development in their units coordination abilities. Equal sharing is a good place to start. Students at Stage 1 can count out a collection of objects and distribute them into a specified number of equal groups. Likewise, they can begin to work with continuous quantities (e.g. rectangular bars) and fragment or partition them into a specified number of equal parts. Chapter 4 will guide teachers through productive interactions involving these kinds of activities.

As students learn to produce equal shares within discrete and continuous quantities, they can begin to consider part–whole relationships. Chapter 5 introduces the *Parts Within Wholes* and *Parts Out of Wholes Fraction Schemes* as a foundation for developing fractions knowledge.

For example, of the latter, $\frac{3}{5}$ can be understood as three equal parts out of five equal parts in the whole. Chapter 5 will guide teachers through productive interactions that build on students' knowledge of equal sharing and toward these foundational conceptions.

Stage 2

At Stage 1, students can develop a Parts Within Wholes Fraction Scheme. When these students take parts out of the whole, the whole is destroyed. For example, with $\frac{3}{5}$, when students take three parts out of five equal parts they do not initially understand that the three parts are still part of the whole. Such an understanding requires that students disembed 3 from 5 – maintaining the three parts contained within the five-part whole even when those three parts are pulled out. This kind of units coordination is associated with a Parts Out of Wholes Fraction Scheme and is available at Stage 2. In addition, students working at Stage 2 can iterate a unit n times to create a composite unit, where the composite unit is understood to be n times as big as the iterated unit. Thus, working at Stage 2 opens possibilities for students to develop more powerful fraction schemes. Chapter 6 focuses on that development.

Beyond part–whole schemes, disembedding and iterating enable Stage 2 students to begin conceptualizing unit fractions as fractional units (i.e. measures). Activities in Chapter 6 focus on ways that teachers can promote measurement conceptions of fractions, especially by engaging students with tasks that require them to iterate unit fractions (e.g. given a $\frac{1}{5}$ piece, ask students to make $\frac{3}{5}$ of the whole or make the whole). We discuss both the critical need for students to engage in such actions and the limitations that Stage 2 students experience in doing so. In particular, students working at Stage 2 treat unit fractions as units of 1 and lose the 1-to-5 multiplicative relationship with the whole when they iterate beyond the whole. Although students working at Stage 2 can iterate $\frac{1}{5}$ seven times to produce $\frac{7}{5}$, they will often name the resulting amount $\frac{7}{7}$.

By treating unit fractions as fractional units, students working at Stage 2 can begin producing and naming proper fractions. However, producing the whole from a given proper fraction requires them to reverse that way of operating. This reversal requires students to coordinate mental actions of partitioning and iterating in new ways: they need to understand partitioning and iterating as inverses of one another. Assessment tasks in Chapter 7 will build on those described in Chapter 6 in order to describe this distinction. Instructional tasks will demonstrate the development that the new understanding affords.

Stage 3

Students working at Stage 3 can take three levels of units as given, so when they iterate unit fractions past the whole to make improper fractions (e.g. $\frac{7}{5}$), they do not lose track of the whole. They maintain the relationships between all three units: $\frac{1}{5}$, the whole and $\frac{7}{5}$. This coordination of units enables students to conceptualize fractions as numbers in their own right and, thus, enables them to engage in a host of new activities. Chapter 8 will specify the tools for assessing this concept and for supporting further development through instruction.

Students at lower stages can begin to understand arithmetical operations with fractions, such as multiplying and adding fractions. They can also begin to conceptualize equivalent fractions and ways to compare the sizes of fractions. However, at Stage 3, students are developmentally ready for instruction that addresses such concepts and their connections in full. Chapters 9–12 will focus on ways that teachers can support those concepts and connections for students at both Stages 2 and 3. For example, activities will help students develop generalizations about multiplying fractions and understand why they can invert and multiply when dividing two fractions.

Chapter 13 extends the reorganization hypothesis into algebraic reasoning. Just as students' ways of operating with whole numbers influence their ways of working with fractions, their ways of working with fractions influence the ways they might work with algebraic expressions and relations.

3

Assessing Stages of Units Coordination

This chapter describes an assessment tool for teachers to use for identifying students' stages of units coordination (Norton, Boyce, Phillips et al., 2015). The tool involves tasks similar to the bars task in Chapter 2 (Figure 2.8). We include sample responses and a rubric for interpreting how students might reason. There are seven tasks in all, but it may not be necessary to use all seven tasks to obtain an initial assessment of their ways of working with units. Rather, each task provides opportunities for students to demonstrate the affordances and limitations of their units coordinating activity. The rubric contains associated indicators for activity at each stage. As described at the end of this chapter, identifying stages in students' units coordinating ability can guide instructional decisions that will support their development of fractions knowledge.

Tasks for Assessing Units Coordination

The first three tasks in the assessment tool are clustered together. The three bars are drawn to scale: 1 unit long, 3 units long and 12 units long. In contrast to the task from Chapter 2, students are not given the relations between the small and medium bars, nor between the medium and long bars. Rather, they must iterate the smaller bars into the longer bars to determine how many times they fit. In fact, they could do the same for the third task, but this would seem a tedious and unnecessary activity for students who readily coordinate two or three levels of units (students at Stages 2 and 3).

Task 4 has the same structure, but this time the bars are not drawn to scale and students must pretend they have the given relations. In this case, iterating the small bar into the long bar will not yield the appropriate relationship. Likewise for Tasks 5 and 6, but in these tasks the unknown relationship is one that requires dividing the numbers in the two given relationships. Task 7 is another reverse task, but now the unknown relationship is not a whole number.

In each of these tasks, students are encouraged to support their responses with drawings, which might provide further indication for how they build and work with units. Teachers can

Name: _____   Teacher: _____   Date: _____

Use the bars shown above to answer the following three questions:

1. How many times does the Medium Yellow Bar fit into the Long Red Bar?

    answer: [   ]

2. How many times does the Small Blue Bar fit into the **Medium Yellow Bar**?

    answer: [   ]

3. Use this information to figure out how many times the **Small Blue Bar** fits into the **Long Red Bar**?

    answer: [   ]

    Use the space below to **draw a picture and explain** your answer.

Name: _____ Teacher: _____ Date: _____

Use the following information to answer questions about the bars shown above:

4. **Pretend** that the Medium Purple Bar fits into the Long Orange Bar *exactly* 2 times.

   Pretend that the Small Green Bar fits into the **Medium Purple Bar** *exactly* 6 times.

   Use this information to figure out how many times the **Small Green Bar** would fit into the **Long Orange Bar**?

   answer: [   ]

   Use the space below to **draw a picture and explain** your answer.

Name: _____   Teacher: _____   Date: _____

Use the following information to answer questions about the bars shown above:

5. **Pretend** that the Medium Purple Bar fits into the Long Orange Bar *exactly* 2 times.

    Pretend that the Small Green Bar fits into the **Long Orange Bar** *exactly* 8 times.

    Use this information to figure out how many times the **Small Green Bar** would fit into the **Medium Purple Bar**?

    answer: [   ]

    Use the space below to **draw a picture and explain** your answer.

Name: _____ Teacher: _____ Date: _____

Use the following information to answer questions about the bars shown above:

6. **Pretend** that the Small Green Bar fits into the Long Orange Bar *exactly* **12** times.

   Pretend that the **Small Green Bar** fits into the Medium Purple Bar *exactly* **3** times.

   Use this information to figure out how many times the **Medium Purple Bar** would fit into the **Long Orange Bar**?

   answer: [   ]

   Use the space below to **draw a picture and explain** your answer.

Name: ———————   Teacher: ———————   Date: ———————

Use the following information to answer questions about the bars shown above:

7. **Pretend** that the Small Green Bar fits into the Long Orange Bar *exactly* 9 times.

    Pretend that the **Small Green Bar** fits into the Medium Purple Bar *exactly* 4 times.

    How can you use this information to figure out how many times the **Medium Purple Bar** would fit into the **Long Orange Bar**?

    answer: ☐

    Use the space below to **draw a picture and explain** your answer.

gather richer evidence by watching students work and probing their reasoning once they respond. However, a teacher should refrain from guiding a student. Questions should focus on understanding the students' reasoning with no assumptions that it should match that of the teacher (e.g. 'how did you know to divide?'). Prompts should be restricted to those that encourage the student to act on her reasoning (e.g. 'try what you were thinking'). During assessment, the goal is for the teacher to learn from the student, not for the student to learn from the teacher.

Rubric for Assessing Units Coordination

Students' actions, including written and verbal responses, provide indicators for how they reason. We cannot see the way they reason, but we can infer it based on those indicators. Rubrics can support these inferences by aligning indicators with particular ways of reasoning. The rubric included here aligns indicators with the three stages of units coordination. It specifically focuses on students' written responses, but as noted previously, if teachers have an opportunity to watch students work and probe their thinking, stronger inferences could be made.

The rubric has four parts: Tasks 1–3 in the first part, Task 4 in the second part, Tasks 5–6 in the third part, and Task 7 in the fourth part. To use the rubric, teachers should review student responses to the tasks and then note which indicators are present in those responses. Student work may include indicators across multiple stages, but if most of the indicators cluster around a particular stage, teachers can make a confident assessment. Note that for the first part (Tasks 1–3) the indicators do not distinguish Stages 2 and 3.

Table 3.1 Rubric for Tasks 1–3

	Student reasoning on Tasks 1-3	Written indicators of reasoning
Stage 1	Students physically or mentally iterate the small bar in the long bar (or segment the long bar with the short bar), relying on the appearance of the bars to determine how many times it would fit.	• Student responses to Tasks 1 and 2 are off by more than 1 from the correct relation (note that students at Stages 2 and 3 might represent the relations as unit fractions). • Student responses to Task 3 indicate that they estimated the number of times the small bar fits into the long bar (possibly further indicated by partitioning marks), rather than taking the product of responses to Tasks 1 and 2. • Students add their solutions to Tasks 1 and 2 to solve Task 3. • Students do not respond, or otherwise indicate they do not know.
Stage 2	Students mentally iterate the medium bar within the long bar four times, with each iteration representing a 3 (i.e. 3, 6, 9, 12).	• Students use relational language (e.g. 'every medium bar is 3 small bars'). • Student drawings incorporate the two relations determined in Tasks 1 and 2. • Student responses justify the use of multiplication.
Stage 3	Students use the given relations to determine that there are four 3s (small bars) in the long bar.	

Table 3.2   Rubric for Task 4

	Student reasoning on Task 4	Written indicators of reasoning
Stage 1	Students rely upon the appearance of the bars without using given relations.	• Students rely upon the appearance of the bars rather than using the given relations (e.g. partitioning/segmenting the given bars). • Students add or subtract the numbers given in the relations. • Students do not respond, or otherwise indicate they do not know.
Stage 2	Students use the second given relation to form a composite unit that they can iterate through activity, by the number in the first given relation.	• Students coordinate relations appropriately and with a drawing illustrating size relations, but writing indicates the drawing was the solution method (e.g. the solution appears below the drawing, or erasures/corrections are present in the drawing). • Student explanations and drawings appropriately refer to multiple two-level relations, but not a single three-level relation. • Student responses indicate use of multiplication without justification or illustration (possibly with a multiplication error).
Stage 3	Students take the first given relation as a composite unit that they mentally distribute across the units given in the second relation, thus justifying the use of multiplication.	• Student drawings are used to justify or illustrate appropriate solutions rather than to produce them (e.g. the drawing is integrated with or appears below an explanation). • Student explanations and drawings refer to a single three-level relation, with appropriate size relations.

Table 3.3   Rubric for Tasks 5 and 6

	Student reasoning on Tasks 5 and 6	Written indicators of reasoning
Stage 1	Students mentally iterate the short (medium) bar, imagining how many times it would fit into the medium (long) bar.	• Students rely upon the appearance of the bars rather than using the given relations (e.g. partitioning/segmenting the given bars). • Students add or subtract the numbers given in the relations. • Students do not respond, or otherwise indicate they do not know.
Stage 2	Students use the two given two-level relations to generate representations with which to relate them figuratively.	• Student drawings or explanations indicate multiplicative reasoning but not reverse multiplicative reasoning (leading them to multiply instead of dividing, possibly because they misread the task). • Student responses indicate use of division, but without justification or supporting illustrations. • Students rely upon their drawings of the given relations to determine the unknown relation. • Student explanations and drawings appropriately refer to multiple two-level relations, but not a single three-level relation.
Stage 3	Students assimilate the two given two-level relations into a structure for coordinating all three levels.	• Students reverse their multiplicative reasoning for both tasks. • Student drawings are used to justify or illustrate appropriate solutions rather than to produce them. • Student explanations and drawings refer to a single three-level relation, with appropriate size relations. • Students use division in ways that are consistent with drawings and explanations.

Table 3.4   Rubric for Task 7

	Student reasoning on Task 7	Written indicators of reasoning
Stage 1	Students mentally iterate the medium bar, imagining how many times it would fit into the long bar.	• Students rely upon the appearance of the bars rather than using the given relations (e.g. partitioning/segmenting the given bars). • Students add or subtract the numbers given in the relations. • Students multiply the numbers in the given relations without any explanation. • Students do not respond, or otherwise indicate they do not know. • Students make no attempt to account for the leftover part.
Stage 2	Students establish a composite unit of 4 and estimate how many of these fit into a length of 9.	• Students refer to fractional part as $\frac{1}{9}$ rather than $\frac{1}{4}$. • Students respond with 2 and a remainder. • Student responses indicate use of division, but without justification or supporting illustrations. • Student drawings or explanations indicate multiplicative reasoning but not reverse multiplicative reasoning (leading them to multiply instead of dividing, possibly because they misread the task). • Student explanations and drawings appropriately refer to multiple two-level relations, but not a single three-level relation.
Stage 3	Students coordinate 9 as two 4s with one unit left over without losing the relationship between this unit and the others.	• Students appropriately account for the leftover part with a fraction or a decimal (e.g. ' $2\frac{1}{4}$ '). • Student drawings are used to justify or illustrate appropriate solutions rather than to produce them. • Student explanations and drawings refer to a single three-level relation, with appropriate size relations. • Students use division in ways that are consistent with drawings and explanations.

Sample Responses

This section provides samples of student work to illustrate how teachers might use the rubric to make inferences about students' units coordinating activity from their written responses. Specifically, we provide examples from student work that fit some of the indicators described in the rubric. We start with sample responses to Tasks 1–3.

For Tasks 1 and 2, students might visually estimate the number of times the blue bar fits into the yellow bar and the number of times the yellow bar fits into the red bar. Reasonable responses could range from 2 to 4 on the first relationship and from 3 to 5 on the second relationship. What is more important than these estimates is what students do with them in responding to Task 3 (determining the number of times the small bar fits into the long bar). If a student does not use the first two relationships to determine this third relationship, or if the student uses them inappropriately, that is an indication of Stage 1 reasoning.

One student responded to Task 3 by reasoning as follows: 'The blue bar goes into the yellow bar 3 times and the yellow bar goes into the red bar 5 times, so you add those to get your answer.' The student had made reasonable estimates of the first two relationships (3 and 5), but he added them to produce the third relationship, indicating that he was not treating the yellow bar as a composite unit containing three blue bars. This strongly indicates Stage 1 reasoning, as described by the following Stage 1 indicator in the rubric (Table 3.1): 'Students add their solutions to Tasks 1 and 2 to solve Task 3.'

Another student included a drawing in his response to Task 3 (see Figure 3.1). The drawing indicates that the student treated each yellow bar as a unit of three blue bars while he was making copies of the yellow bar to produce the red bar. We can infer from this activity that the student was working with a composite unit. This activity fits the following Stage 2/3 indicator: 'Student drawings incorporate the two relations determined in Tasks 1 and 2.'

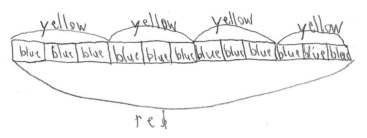

Figure 3.1   Student drawing in response to Task 3

Beginning with Task 4, the bars in the tasks are not drawn to scale, which provides opportunities for teachers to begin distinguishing Stage 2 and Stage 3 reasoning. In Task 4 students are asked to pretend that the medium bar fits into the long bar two times and that the small bar fits into the medium bar six times; they are asked to use these given relationships to determine the relationship between the small bar and the long bar. Indicators of Stage 3 reasoning suggest that the student can assimilate all three levels of units in the tasks without having to build up the third level through some kind of activity. One way to make such a distinction is to consider student drawings: do student drawings show the coordination of multiple two-level relations or do they show coordination of all three levels in one drawing?

Consider the response illustrated in Figure 3.2. Although the answer, 12, is correct, the drawing indicates that the student coordinated only two levels of units at a time. Once she had represented the relationship between the small (green) bar and the medium (purple) bar, she could use it to build the relationship between the small (green) bar and the long (orange) bar, but there is no indication that she had assimilated all three levels of units at once. Her response is consistent with the Stage 2 indicator, 'Student explanations and drawings appropriately refer to multiple two-level relations, but not a single three-level relation' (Table 3.2).

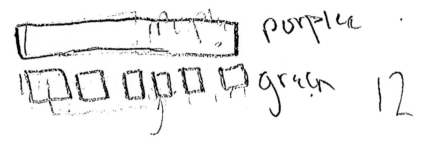

Figure 3.2   Drawing depicting a two-level relationship

In contrast, the response illustrated in Figure 3.3 indicates that the student had all three levels of units in mind at the same time: the long bar is drawn as both 12 small bars and as two medium bars, each of which contains six small bars. This fits with the Stage 3 indicator, 'Student explanations and drawings refer to a single three-level relation, with appropriate size relations' (Table 3.2).

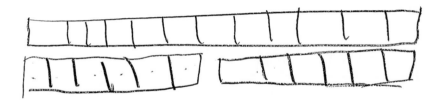

Figure 3.3   Drawing depicting a three-level relationship

Tasks 5 and 6 require students to work in reverse. Students are given the relationship between the small bar and the long bar, as well as one of the other relationships. Rather than building up the long bar from the small and medium bars, students need to determine an unknown intermediate relationship. In Task 5, the unknown relationship is between the small bar and the medium bar, given that the small bar fits into the long bar 8 times and that the medium bar fits into the long bar 2 times. Students at Stage 2 may have to determine this relationship by relying on figurative material. For example, consider the student work shown in Figure 3.4.

The student represented the two given relationships in his picture: eight small bars in the long bar and two medium bars in the long bar. Then he relied on this picture to estimate the number of small bars in the medium bars, responding 'about $3\frac{1}{2}$ times' (though he wrote '$\frac{1}{3}$' in his picture). This reasoning fits the Stage 2 indicator, 'Students rely upon their drawings of the given relations to determine the unknown relation' (Table 3.3).

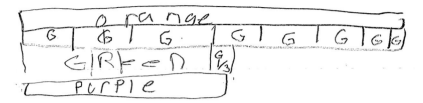

Figure 3.4   Response indicating Stage 2 reasoning on Task 5

The student response shown in Figure 3.5 (responding to Task 6) indicates Stage 3 reasoning. In this task the unknown relationship was between the medium bar and the long bar, given that the small bar fits into the long bar 12 times and fits into the medium bar 3 times. The student claimed, 'you do 12 divided by 3', and her work is supported by her drawing. She represented all three relationships in her picture, which is consistent with her calculation. This response fits with all four Stage 3 indicators (Table 3.3), particularly the following:

- 'Student explanations and drawings refer to a single three-level relation, with appropriate size relations.'
- 'Students use division in ways that are consistent with drawings and explanations.'

Figure 3.5   Response indicating Stage 3 reasoning on Task 6

Task 7 introduces an additional complication in that the unknown relation is not a whole number. Thus, there is another unit to consider – the leftover unit fraction. Students are told that the small bar fits into the long bar 9 times and fits into the medium bar 4 times; then they are asked to determine the number of times the medium bar fits into the long bar. Because students at Stage 2 have to build up the third level of units through activity, they will have trouble dealing with this extra unit, which represents a fourth level. For example, Stage 2 students might misname the fractional unit (e.g. $\frac{1}{9}$ rather than $\frac{1}{4}$), ignore it, or avoid it altogether by multiplying the two given relations rather than dividing them.

Students at Stage 3 can assimilate the first three levels in the task, so they can devote more attention to the fractional unit. Although they might also misname the unit fraction, they are less likely to ignore or avoid it. Figure 3.6 illustrates two responses from Stage 3 students.

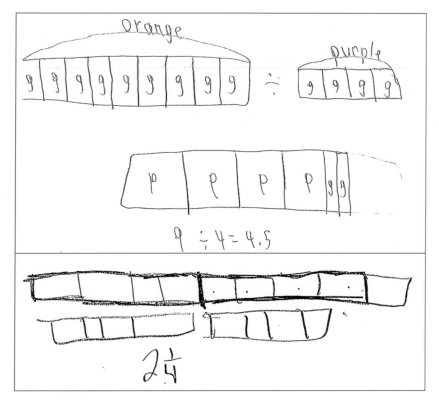

Figure 3.6 Response indicating Stage 3 reasoning on Task 7

The response at the top of the figure is incorrect, but the student does seem to take all of the units into account. In the process, he seems to have confused the number of times the small (green) bar goes into the medium (purple) bar with the whole number of times the medium bar fits into the long (orange) bar. Thus, he answers '4.5' rather than '$2\frac{1}{4}$', and his drawing is consistent with this answer. The response at the bottom of the figure demonstrates the correct answer with a corresponding drawing. We can be more confident in this response as indicative of Stage 3 reasoning; however, like assessment in general, assessing students' stages of units coordination is not a simple matter of judging correctness.

Having assessed students' stages of working with units, teachers are better positioned to support students' development of fractions knowledge. Chapters 4 through 12 are organized around these stages: Chapters 4 and 5 build upon Stage 1 reasoning; Chapters 6 and 7 build on Stage 2 reasoning; and Chapter 8 builds on Stage 3 reasoning. Chapters 9–12 focus on arithmetic with fractions (e.g. adding fractions, dividing fractions) building on Stages 2 and 3. Each of these chapters contains fractions-specific assessments for students' reasoning within each stage. Chapter 13 outlines relationships between units coordination, fractions knowledge and algebraic reasoning.

Activities contained in these chapters should also help promote students' development to higher stages of units coordination. Although students begin to develop units coordination in whole number contexts, they can continue to develop this ability in fractions contexts (Boyce and Norton, 2016). The general instructional approach is to engage students in activities that require them to coordinate additional levels of units through activity (Norton and Boyce, 2015). We note that, as an additional resource for promoting units coordination, there is a free app for the iPad on iTunes – *CandyDepot* – that was designed for that purpose (for more information on this app see http://ltrg.centers.vt.edu, or Norton, Boyce and Hatch, 2015).

4

Teaching Students at Stage 1: Fragmenting

The word *fraction* comes from words that mean to fragment or break apart (Merriam-Webster online, 2015). For young students, the origin for fragmenting can come accidentally, like dropping a plate on the floor, or intentionally, such as breaking a cookie into parts to share with a parent or sibling. A critical psychological foundation for making fractions is *equally* sharing 'stuff' that can be fragmented, or subdivided (Empson and Levi, 2011; Piaget et al., 1960; Steffe and Olive, 2010: 58; Streefland, 1991). How a student makes equal shares evolves significantly in the elementary school years, and students in the same class may demonstrate quite a variety of approaches to equal sharing problems.

Imagine a young child, Lila, who has a clay rectangular cake and aims to share it with herself and some action figures. Making equal shares means breaking or marking the cake into parts, which we refer to as *fragmenting*. Children demonstrate fragmenting at different levels of sophistication (Steffe, 2010a: 68–70), and the number of equal shares to make can greatly affect the difficulty of making the shares equal. For example, if there are four action figures (so five shares to be made), Lila may pay close attention to making five equal parts and not use up all of the cake. Alternately, Lila may use up all of the cake but not make equal parts. The extent to which the child is bothered by not using all of the cake, or by not making equal parts, is one important indicator of the level of sophistication of their fragmenting activity.

Trying to equally share a cake provokes an important question: how can we tell whether the shares are equal? If a child like Lila uses up all of the cake but does not make equal parts, someone (an adult, the child, a child's friend or sibling) might wonder whether the shares are equal, or if not, whether any of the parts represents an equal share of the cake. Many students respond by stacking or lining up the shares to assess equality. A more advanced activity involves replicating one of the shares five times to see whether five of these are the same as the whole cake. If so, then that share represents an equal share and others could be 'adjusted' to that size.

Successful equal sharing of single items can allow students to develop fraction language and fraction notation that are linked to the fragmenting they do to equally share. Unfortunately, for many older students fraction language and fraction notation are divorced from fragmenting (e.g. Olive and Vomvoridi, 2006), which means these students have drastically different responses to a request to share a cake equally among five people and a request to draw one-fifth of the cake.

This chapter focuses on the origins of making fractional parts through equal sharing. We introduce five levels of fragmenting and describe how those are related to the stages from Chapter 3. We also identify ways to work with students at the first three levels of fragmenting.

Fragmenting and Partitioning

There are five different levels of fragmenting that elementary school students may demonstrate – perhaps all in the same classroom (Steffe, 2010a: 68–70). These levels were developed from research on ways of thinking that are necessary to produce fractions (Piaget et al., 1960; Steffe and Olive, 2010). In particular, Piaget and colleagues found that to build the idea of a fraction students must (1) conserve the whole, which means being able to break up a whole and reassemble it from the parts; (2) be aware that parts should be equal in size; (3) use an appropriate number of parts (e.g. if making equal shares for five people, one would make five parts); and (4) see the parts in relation to each other and in relation to the whole of which they are a part. This latter way of thinking is sometimes expressed as seeing part-to-part and part-to-whole relations.

The five levels of fragmenting (Steffe and Olive, 2010) incorporate these ideas from Piaget and colleagues (Piaget et al., 1960). The levels also incorporate connections to students' numerical knowledge, because the ways that they have organized discrete units and composite units (units of units) affect how they use those units to mark lengths into parts. For example, some pre-numerical students can mark a length into two or three parts but are unlikely to mark it into five or more parts. Furthermore, those parts may not be equal. These students demonstrate the *first level of fragmenting*. By pre-numerical we mean that they are in the stages of Emergent Counting, Perceptual Counting or Figurative Counting as laid out in the Learning Framework in Number (Wright et al., 2006a).

Children who partition similarly to Lila in the opening of this chapter demonstrate the *second level of fragmenting*. These students are challenged to coordinate their two goals of making equal parts and using up all of the material: a focus on one goal tends to preclude attention to the other. Some students like Lila might eventually achieve a successful coordination after many trials. The second level of fragmenting corresponds to having constructed the Initial Number Sequence (Steffe, 2010b); the Initial Number Sequence indicates that a child is numerical (see Biddlecomb, 2002; Steffe, 1994; Wright, 1994; Wright et al., 2006a).

When students regularly coordinate both goals (to make equal parts and to use up all of the material) with three parts, they demonstrate the *third level of fragmenting*. These students may coordinate both goals with a higher number of parts as well. They often use experimentation to achieve the coordination with a higher number of parts. The third level of fragmenting corresponds with students operating at Stage 1, described in Chapter 3. At this level we start to use the word 'partitioning' to describe what students are doing. We view partitioning as a special kind of fragmenting in which students regularly coordinate both goals (to make equal parts and to use all of the material) and are starting to develop ways to justify whether the shares are equal.

At the *fourth level of fragmenting*, students coordinate both goals for any number of parts (within reason). The fourth level of fragmenting corresponds with students operating at Stage 2, described in Chapter 3. These students understand that any equal part of a whole could be repeated the requisite number of times to recreate the whole. So, if a part repeated the requisite number of times does not recreate the whole, it is not an equal share. For example, let's say that Dewayne is sharing the clay cake equally among himself and four action figures. He makes marks for the five shares (Figure 4.1, *top*). If Dewayne or someone else questioned whether the parts were equal, one way Dewayne could test would be to repeat a part five times to see if it recreates the whole. For example, the lower rectangle in Figure 4.1 represents Dewayne choosing the second part in the top bar and repeating it five times to see if that produced the original whole. The activity of Dewayne and other students like him can be called equi-partitioning.

Figure 4.1 Equi-partitioning

The *fifth level of fragmenting* involves sharing multiple items equally among multiple people, for example sharing four bars equally among six people. Students operating at the third and fourth level can certainly engage in these kinds of equal sharing problems. Students who solve these kinds of equal sharing problems in a particular way demonstrate the fifth level of fragmenting, and this level corresponds with students operating at Stage 3. We address these problems and students' solutions in Chapter 9. The stages and levels of fragmenting are summarized in Table 4.1.

Table 4.1 Stages and levels of fragmenting

Stage (from Chapter 3)	Level(s) of fragmenting	Characteristics
0	1 and 2	Mark a length into two or three parts, not more, and the parts are unlikely to be equal. Find it challenging to coordinate the two goals of making equal parts and using up all of the material.
1	3	Regularly coordinate two goals of making equal parts and using up all of the material with three parts, and may do so experimentally with greater numbers of parts.

(Continued)

Table 4.1 (Continued)

Stage (from Chapter 3)	Level(s) of fragmenting	Characteristics
2	4	Coordinate two goals of making equal parts and using up all of the material for any number of parts (within reason); understand that any equal part of a whole could be repeated the requisite number of times to recreate the whole.
3	5	Can share multiple items equally among multiple people in ways that involve distributive reasoning, addressed in Chapter 9.

Testing Equality of Shares

Asking questions like 'how do you know the shares are equal?' is an important prompt for students who are engaged in equal sharing problems. At the first level of fragmenting, students may not find the question that meaningful. If they do, they will likely respond by cutting apart the shares and lining them up (Figure 4.2) or placing them on top of each other to assess overlap. Students at the second and third levels of fragmenting are also quite likely to act in this way.

Figure 4.2   Separating the parts and lining them up to assess equality

The foundation for this way of acting is a way of thinking: parts are equal if they match exactly. Although this way of thinking is critical, it is not sufficient for developing fractions knowledge because the idea behind making, say, five equal parts in a whole is that any one of those parts could be repeated five times to make the whole. That way of thinking opens the way for $\frac{1}{5}$ to be repeated five times to make 1, and it is not yet available to students at the first three levels of fragmenting.

Students at the fourth and fifth levels of fragmenting can develop this other way to test equality of shares – but doing so is not automatic: they must have opportunities to develop this way of thinking. The question 'how do you know the shares are equal?' alone may not be a sufficient prompt. In addition, it is important to single out one share and ask, 'How do you know if this part is an equal share?' For example, let's say Patrice had attempted to make five equal shares as shown in Figure 4.3. A teacher asks Patrice whether the shares are equal and she immediately responds that they are not. A good follow-up question is: 'Do you think any of them is an equal share for five people, and how do you know?' If students are stumped by this question, it can be helpful to be more suggestive about the action they might take. For example: 'Could you use this part (pointing to a part, not one that will be an equal share) to mark off parts on this bar (another bar of the same size) to find out if it is an equal share?'

Ask: 'Do you think any of them is an equal share for five people, and how do you know?' Or 'Could you use the first part (pointing) to mark off parts on this bar (another bar of the same size) to find out if it is an equal share?'

Figure 4.3   Making equal shares

Asking questions about a particular share gives students an opportunity to develop what we refer to as 'iterating', or repeating a part some number of times for a purpose, such as testing equality or making a larger fraction from a smaller one. The questions about a particular share also give students an opportunity to develop what we refer to as 'disembedding', or taking a part out of a whole while mentally keeping the whole intact. Students at the first or second levels of fragmenting are unlikely to demonstrate iterating or disembedding (Steffe and Olive, 2010). Students at the third level of fragmenting can learn to iterate, but they do not yet disembed (Hackenberg, 2013; Steffe and Olive, 2010). Students at the fourth and fifth levels of fragmenting can demonstrate both iterating and disembedding. In this chapter we are focused on students operating at the first three levels of fragmenting; we address students at the fourth and fifth levels in later chapters.

Working with Students at the First Three Levels of Fragmenting

Our general recommendation for students at the first three levels of fragmenting is to help them correlate their work in fragmenting with their developing whole number knowledge. There are three broad types of problems that are useful toward this end. We note that these problems can be useful for students at the fourth and fifth levels as well, but the responses of these students will be different from the responses of students at the first three levels.

Fragmenting Problem Type 1: Equal Sharing

The first type of problem is to ask students to share a segment or area (we recommend rectangular strips) equally among a number of people and justify whether they have made equal shares. Good questions to ask include these: 'The people are sharing fairly; did everyone get the same size piece? How do you know?' We have already discussed student responses to this type of task in the prior two sections of this chapter.

Here we add that these tasks must involve the teacher in introducing fraction language in order for students to link equal sharing experiences, and so fragmenting and partitioning, to fraction language. For example, if students share a rectangular sub sandwich equally among three people and justify that they have made equal shares, teachers can let them know we can call one share 'one of three (equal) shares'. Then as students continue to work on these problems, teachers can use the phrase 'one share out of three (equal) shares'. The word 'equal' is in parentheses because, depending on the context, it might be okay sometimes to omit it. We recommend starting with these ways of phrasing number words because these phrases refer to the action of taking one part out of three equal parts. We advise against using 'one-third' with students at the first three levels of fragmenting because this number word describes no obvious action.

Once students are familiar with equal sharing tasks and naming equal shares with phrases like 'one share out of three (equal) shares', teachers can introduce written notation like 1/3 and say it means 'one out of three (equal) parts'. We note that this pathway of introducing fraction language and symbols highlights two of the eleven dimensions of mathematizing, **notating** and **formalizing**. This pathway represents **notating** because it is faithful to students' ways of thinking and highlights the relation, or comparison, between the part and the whole, which is essential for building fraction knowledge. It also represents **formalizing** because the teacher is introducing formal symbols like $\frac{1}{3}$.

Fragmenting Problem Type 2: Drawing Unit Fractions

Once students start to talk about their equal sharing experiences with fraction language (and possibly with written notation as well), then teachers can pose problems that involve fraction language. The second type of problem is to ask students to draw a unit fraction of a segment or area and justify whether the result is that fraction.

Consider this problem:

'Drawing a Unit Fraction Problem': Here is a carrot cake (a rectangle, or bar). Molly's family is going to eat [one out of two, one out of three, one out of four, one out of five] equal parts of the carrot cake.

a)   Can you draw [one out of two, one out of three, one out of four, one out of five] equal parts of the carrot cake?

b)   How do you know that amount is [one out of two, one out of three, one out of four, one out of five] equal parts of the carrot cake?

For students at the first level of fragmenting, one out of two equal parts would be a good choice, because students' knowledge of one-half often runs ahead of their knowledge of other fractions (Behr et al., 1984; Hunting and Davis, 1991). We expect that students at the first level of fragmenting may make a mark on the bar to make two parts, but these two parts are unlikely to be equal. Alternately, they may make two marks on the bar (yielding three parts) because 'one out of two equal parts' brings up 'two' for them. These students may not feel any logical necessity to investigate equality of parts in trying to justify that they have made one out of two parts of the bar.

For students at the second level of fragmenting, one out of two parts or one out of three parts would be good choices (Biddlecomb, 2002). Let's choose one out of two parts. Students at this level are likely to draw a bar that is a little bigger or smaller than one-half of the bar (Figure 4.4).

Students at the second level of fragmenting respond to the problem of drawing one out of two equal parts of the given carrot cake

Figure 4.4 Attempting to draw an equal part

When a teacher asks how the students know that amount is one out of two parts, the students might use their part to mark off a part on the original bar. Then they could fold the bar to make a judgment (Figure 4.5).

Figure 4.5 Marking the original bar and folding it

If the two resulting parts of the original bar are not equal, the students may state that their estimate was one out of two equal parts of the cake and also claim that the remaining part in the original bar (which is smaller than their part) is one out of two equal parts of the cake (Biddlecomb, 2002). So, these students produce an appropriate number of pieces and know that making equal shares involves breaking up and reassembling a whole. That is, they know that they should conserve the whole. However, they do not demonstrate a need to make equal parts. Teachers can ask these students about whether they think both parts need to be equal. We caution that some students at this level will say that the parts should be equal while also continuing to act in the way we have just described.

Alternately, if these students have made an estimate where two of their parts together are smaller than the whole cake, the students may mark two of their parts on the original and cut off the excess on the cake (Figure 4.6).

Excess that students may cut off

Figure 4.6 Marking the original bar twice and cutting off the excess

In this case the students produce an appropriate number of pieces and have made two equal parts. However, they have not conserved the whole. Teachers can ask these students if they can make one out of two equal parts of the whole original cake: That is, teachers can tell them that they did a nice job, but now to try it again without cutting off part of the cake.

For students at the third level of fragmenting, any of the number choices could be used. In contrast with students who are at the second level of fragmenting, these students are more likely to adjust their estimate if it is too big or too small. However, because they have not yet constructed disembedding, they cannot conceptually take a part out and keep the whole intact. So, once they have partitioned a bar and broken it into the parts, they do not conceptually use those parts to recreate the whole. In other words, their reasoning is consumed by the mental action of partitioning. So, they cannot stand back and see the results of that mental action as parts that can be used in further reasoning, while maintaining the relationship of those parts to the whole (Olive and Vomvoridi, 2006).

We also note that at any of these three levels, teachers can always return to equal sharing problems to try to develop fragmenting and partitioning. Hackenberg (2013) has posed the 'Drawing a Unit Fraction Problem' with one of three parts to a 7th-grade student who took 'three little pieces' off the cake and indicated that the remaining amount of cake was $\frac{1}{3}$ (Figure 4.7). That is, Courtney's drawing of one-third was, from our perspective, more than $\frac{3}{4}$ of the given bar.

Figure 4.7 Seventh-grade student Courtney's drawing of one-third of a given cake

In this case, it is important for teachers to go back to the sharing language: 'Okay, here's another copy of the cake. How can you share it equally with three friends?' Many students at the second and third level of fragmenting will find this problem more sensible than a request to draw a fractional amount. However, again, once the students make the equal shares, teachers can continue to help them give a fraction name for one of the shares ('one of three parts' or 'one out of three parts') and to write fraction notation for one of the shares.

Fragmenting Problem Type 3: Making Wholes from Parts

The third type of problem is to ask students to iterate a part to make a whole. In our view this problem type is more useful when students are beyond the first level of fragmenting.
    Consider this problem:

'Making the Whole Problem': Here is an entire zucchini cake and here are parts that are one out of two, one out of three, one out of four, one out of eight, and one out of 12 equal parts of the cake (Figure 4.8).

a)   Can you use the _____ [teacher selects one part] to make the whole cake?

b)   How did you know how many of those parts to use?

    (adapted from Biddlecomb, 2002)

Figure 4.8   Picture for the 'Making the Whole Problem'

For students at the second and third levels of fragmenting, one out of two, one out of three and one out of four are good choices, at least initially. These students are likely to repeat the part the requisite number of times to make the whole zucchini cake, although they may be using visual comparison to determine the number of parts they need. As the students encounter more of these problems, teachers can try to use smaller parts. Overall, the goal of these problems is to promote the view that if a fraction is, say, one of four parts of the whole, then four of those parts must make the whole. Although this may seem obvious to teachers, it is an achievement in thinking for students at the second and third levels of fragmenting.

Assessment Task Groups

List of Assessment Task Groups

A4.1: Equal Sharing of Single Items

A4.2: Making One Equal Share

A4.3: Drawing a Unit Fraction of One Bar and Justifying the Amount

A4.4: Making Connected Numbers

A4.5: Drawing a Unit Fraction of a Segmented Bar and Justifying the Amount (adapted from Biddlecomb, 2002)

Task Group A4.1

Equal Sharing of Single Items

Materials: A variety of rectangles made from cardboard, with multiple copies of the same size; writing utensil for making marks; scissors for cutting bars.

What to do and say: Ask the students to share one of the rectangles (it can represent a granola bar or a healthy cake) equally among some number of people. Ask the students to justify their result: How do they know they have made equal shares? Can they justify in more than one way? Do the students think any of the parts are an equal share for that number of people, and how do the students know?

Notes

- Starting with two and three equal shares is important when working with students at the first, second, and third levels of fragmenting.
- If a student does not create more than two equal shares when asked for more than two, that is an indication that the student is at the first level of fragmenting.
- If a student makes equal shares or exhausts the whole rectangle (with no leftover piece), but does not accomplish both goals together, that is an indication that the student is at the second level of fragmenting.
- If a student makes equal shares for smaller numbers (2, 3, or 4) but struggles to make equal shares for larger numbers (5–10), that is an indication that the student is at the third level of fragmenting.
- Students who regularly create equal shares while exhausting the whole, and who can check their work through iterating (repeating) one part, are likely at the fourth or fifth level of fragmenting.

Task Group A4.2

Making One Equal Share

Materials: A variety of rectangles made from cardboard, with multiple copies of the same size; writing utensil for making marks; scissors for cutting bars.

What to do and say: Ask the students to share one of the rectangles equally among themselves and two friends, so three people in all. However, the students are to make only one mark to mark off their own share. Ask the students to assess and test their result: Do they think they have made an equal share if three people are sharing? How can they tell? How can they justify it to someone else?

Notes

- This task is difficult for Stage 1 students because to be fully successful requires a disembedding operation. However, that means it can be a good task for assessing the ways Stage 1 students coordinate the two units (the part and the whole) through their activity.
- Students at the third level of fragmenting may be able to solve the task by guessing an initial mark and then imagining marking off that length on the rectangle (i.e. iterating that length), possibly using their fingers to keep track.
- Students at the fourth and fifth levels can make accurate estimates without much activity, although they might feel the need to check their estimates by imaginatively iterating the part.

Task Group A4.3

Drawing a Unit Fraction of One Bar and Justifying the Amount

Materials: Unmarked fractions strips (long, thin, rectangular strips of paper; see Appendix), with several copies of the same size rectangle; writing utensils for making marks.

What to do and say: Tell the students that the rectangles represent loaves of banana bread. Ask the students to draw a unit fraction, such as one out of two equal parts, of a specified loaf. How do the students know that what they drew is one out of two equal parts? Can they justify in more than one way? Repeat with other unit fractions.

Notes

- If students are successful with one of two equal parts, try one of three equal parts and one of four equal parts.
- This task is especially appropriate for assessing whether students can operate at the third level of fragmenting. Students at this level can learn to iterate the unit fractional part to check whether (1) it exhausts the whole and (2) it goes into the whole the right number of times.

Task Group A4.4

Making Connected Numbers

Materials: A small thin rectangle made from thick cardboard or plastic, paper, writing utensils.

What to do and say: Ask the students to trace the rectangle on the paper and call that 1 length unit, a 1-stick. Ask them if they can create a length that is two times the 1-stick. If they suggest tracing another rectangle,

(Continued)

(Continued)

ask them to do it and say that we'll call that length a 2-stick. Ask them if they can make a length that is three times the 1-stick. If they make a length that is three of the rectangles, say that we'll call that a 3-stick. Ask them if they can make a 4-stick, and can they do it in more than one way. They could join a 3-stick and 1-stick, or copy the 2-stick two times. If they repeat the 1-stick four times, that could be a sign of iterating.

Notes

- Teachers can pose similar tasks using the free online applet *JavaBars*, which can be found on John Olive's web page (http://math.coe.uga.edu/olive/welcome.html).
- In this assessment activity students are making what we refer to as 'connected numbers' – a sequence of continuous wholes created by iterating a continuous unit so many times. Doing so is important because it allows students to reorganize their concepts of discrete whole numbers to create a set of lengths, an early precursor of a number line (Steffe and Olive, 2010).
- Students at the first three levels of fragmenting may not make the 2-stick separately from the 1-stick. That is, they may join one more traced unit onto the unit they have already traced. One reason that they might do this is that they are not disembedding the original unit as its own length in relation to the length that is twice as long. If that occurs, it is important to note it as part of the assessment. However, to get the task going you can say something like: 'Oh, the 1-stick is not on our paper separately anymore. Can you draw the 1-stick again, so we have the 1-stick and the 2-stick?' If that is still not sensible to the students, you can draw the 1-stick separately and see if the students find this to be sensible.
- Many students at Stage 1 will make connected numbers by joining a 1-unit onto the previous stick. That's certainly okay. However, asking them if there is more than one way to make the target stick can open up more possibilities for them.

Task Group A4.5

Drawing a Unit Fraction of a Segmented Bar and Justifying the Amount (adapted from Biddlecomb, 2002)

Materials: Multiple cardboard rectangles that are partitioned into some number of equal parts greater than six, such as eight equal parts or twelve equal parts. We will refer to these rectangles as 8-bars or 12-bars. Also needed are writing utensils and extra paper.

What to do and say: Ask the students to draw a bar that is one of two equal parts of the 8-bar. How do they know that is one of two equal parts of the 8-bar? Can they make one of two equal parts of the 8-bar in another way? How do they know that the new bar is one of two equal parts of the 8-bar?

Notes

- Students may draw a copy of the bar that is unmarked, place a mark on the bar and then transfer it to the 8-bar (Biddlecomb, 2002). If so, this is an indication that the student does not think of the bar as consisting of eight parts that can be subdivided.
- If students do shade or draw out four parts of the 8-bar, that indicates that they are coordinating their whole number knowledge (4 is one of two equal parts of 8) with their ideas about parts and that they are thinking of the 8-bar as eight parts that can be subdivided.

- Students may make one of two equal parts in another way by shading in a different four parts on the 8-bar. If they justify their work by saying that they have made 4, that is a sign that they can take any of the four parts and think of it as a unit in relation to the 8-bar. Indeed, this can be a sign that a student at the first or second level of fragmenting is producing a composite unit (Biddlecomb, 2002).

Instructional Activities

List of Instructional Activities

IA4.1:   Share an Energy Bar Fairly

IA4.2:   Make Only Your Own Share

IA4.3:   Slicing Cakes in a Bakery

IA4.4:   Working with Connected Numbers

Activity IA4.1

Share an Energy Bar Fairly

Intended learning: Students will learn to make and justify equal shares, thereby progressing along their path of fragmenting and partitioning.

Instructional mode: Students working in pairs with instruction from the teacher.

Materials: A variety of rectangles made from cardboard, with multiple copies of the same size; writing utensil for making marks; scissors for cutting bars; a die or spinner with 1–5 or 1–6 on it.

Description: Working in pairs, students will be asked to select two copies of a rectangle that represents their energy bar. One student rolls the die or spins to see how many people they will share their bar with. When they have a number, each student works separately to make the equal shares. When they each have completed their work, they show each other and decide together who has made equal shares *and* used up all of the energy bar (could be both students, neither student, or one of the students). If they think that one bar has not been equally shared, they are both to respond to the question: 'Do you think any of the parts is an equal share for ___ people, and how do you know?' If they think that one bar has been equally shared, they are to show how they know that and name a single part. An important part of this activity is the teacher circulating in order to hear discussions about how students know that they have made equal shares. So, some time should also be spent having a whole class discussion about the justifications that students have made.

Responses, Variations and Extensions

- If students do not use up all of the rectangle, the teacher can ask them to try again but share the whole energy bar, thereby helping them to understand that the goal is to share the whole rectangle equally.

(Continued)

(Continued)

- Many students can make two equal parts (halves) but struggle with larger numbers for partitioning. Through this activity, students should begin to extend the range of numbers with which they can partition.
- This activity also connects to geometry standards related to partitioning areas, such as Common Core State Standards 2.G.3 in the United States.

Activity IA4.2

Make Only Your Own Share

Intended learning: Students will make a single mark on a bar to show their own share when sharing the bar equally among some number of people, which can promote iterating and disembedding.

Instructional mode: Game that can be played with partners.

Materials: A variety of rectangles made from cardboard, with multiple copies of the same size; writing utensil for making marks; scissors for cutting bars; die or spinner with 1–5 or 1–6 on it.

Description: Working in pairs, students will select two copies of a rectangle that represents their energy bar. The teacher rolls the die (or spins) to determine how many people are to share the bar in the first round. So, in this game, everyone in the class is using the same number of people, but not necessarily the same size bar (partners should have the same size bar). The teacher tells students that if he rolls a 3, that means three students in all share the bar. Students are to make *one* mark on their bar to show one of the equal shares. Then the partners compare their bars and decide who came the closest to making an equal share. To make this decision, students may decide to take a variety of actions, such as cutting the part off of the bar, folding the bar, etc. When all partners have made their decisions, the teacher asks for a report to the class. Each partner reports out whether they had an equal share and how they know. The teacher can lead students in naming equal shares with a fraction name (e.g. 'one of three equal parts') and fraction notation. Students can earn points for making an equal share, and partners can keep track of their points after each new round of the game.

Responses, Variations and Extensions

- Students at Stage 1 may begin to iterate and disembed by relying on figurative material (e.g. cardboard cutouts or drawings). Teachers should allow these students plenty of time to experiment with these materials in order to check their initial responses. The goal is for students to internalize those actions so that they become mental actions. This internalization is supported by gradually distancing the instructional setting as students rely less and less on the instructional materials.
- If there are different size bars being used in the class and more than one pair of students successfully makes an equal share, then students will produce an equal share for three people from a larger bar, and an equal share for three people from a smaller bar. The teacher can take this opportunity to ask why we are calling two different size parts $\frac{1}{3}$. Students may find it odd. To help with the discussion, the teacher could ask if there are different sizes of '5'. Students might relate to the idea that 5 centimetres and 5 miles are very different but are both 5. So, the size of a number depends on what we have decided '1' means.

- Naming and **notating** the unit fractional parts marks an entry point to **formalizing** fractions and meeting early fractions standards, such as Common Core State Standard 3.NF.1.

Activity IA4.3

Slicing Cakes in a Bakery

Intended learning: Students will learn to make unit fractions of wholes, linking up their partitioning activity with fraction language. Students will also learn to justify their responses, which may involve some students in iterating.

Instructional mode: Students working in pairs with instruction from the teacher.

Materials: Multiple copies of the same size rectangles made from cardboard; writing utensils; scissors.

Description: The teacher will tell students that they are all workers in a bakery that sells delicious cakes that are in a special rectangular shape. Customers come into the bakery and order a slice with a fraction name (unit fractions only). So, a customer might come in and say: 'I want one of two equal parts of the cake.' Students are to fill that order by slicing the cake and cutting the appropriate amount. After the students have cut a slice, they must justify to the customer that they have filled the order correctly. At first the teacher is the customer, and all the students work on the same unit fraction. That way cake slices can be compared across the class, and discussions about justifications can be easily shared. As the students become used to the activity, variations and extensions can be followed as described below.

Responses, Variations and Extensions

- Students can be partnered and use different size bars, as in Activity IA4.1 and Activity IA4.2.
- Students could keep track of points as in Activity IA4.2; that is, students could earn a point (or pretend coin) each time they correctly fill a customer's order.
- If students struggle with this activity, it is best to return to equal sharing language and activities, as in Activity IA4.1 and Activity IA4.2.
- As noted in Chapter 2, iterating unit fractions marks significant progress in **complexifying** fractions (from simple part–whole relationships toward measurement concepts). It also builds on students' use of iterating units of measure (e.g. centimetres) as addressed in early standards, such as Common Core State Standards 1.MD and 2.MD.

Activity IA4.4

Working with Connected Numbers

Intended learning: Students will create connected numbers and reason with them.

Instructional mode: Students working in pairs with instruction from the teacher.

Materials: Short, thin rectangles made from thick cardboard or plastic, paper, writing utensils.

(Continued)

(Continued)

Description: Students start by selecting a small bar that will be their unit. Teachers pose the task of making a set of bars that are twice as long, three times as long, etc., up to ten times as long, as the unit. Students can make the bars by copying the unit and joining the copies together, but some students will start to copy and repeat the unit – a possible sign of an iterating operation. When the students have completed their set of bars, the teachers will pose questions about all the different ways to make bars. For example, the teachers can ask how many ways students can make a 5-bar (a bar five times as long as the unit). Responses include iterating the unit five times, or joining a 2-bar and a 3-bar. The teachers will also pose tasks about bars beyond the 10-bar that students could make. For example, the teachers can ask what a bar three times longer than an 8-bar will be. Students will make the bar and explain how they did it.

Responses, Variations and Extensions

- Like Task Group A4.4, these tasks can be posed using the *JavaBars* applet, see A4.4.
- The goal of this task is for students to think about numbers as measures and to understand how these numbers are made up of other numbers/measures. Thus, it is important that students have time to find multiple ways of making the various connected numbers. Some students will come up with more ways than others.
- If some students need a greater challenge, the teacher can ask them how they know whether they have found *all* of the ways to make the connected number.

5

Transitioning to Stage 2: Part–Whole Reasoning

Domain Overview

One of the most common responses to the question 'What is a fraction?' is 'It's a part of a whole'. Indeed, the idea that fractions are parts of wholes is basic, pervasive and has been discussed for quite some time (Kieren, 1980; Piaget et al., 1960). The authors of this book, along with many others, advocate that thinking of fractions as only parts of wholes limits building robust fractions knowledge (e.g. Thompson and Saldanha, 2003). However, children begin with ideas of fractions as parts within wholes or parts out of wholes, and it is important to understand that these ideas vary across students operating at Stages 1, 2 and 3.

The idea that a fraction is a part in relation to a whole never goes away; it will always be a part of a person's concepts of fractions as they mature, so it is fundamental (Piaget et al., 1960). This chapter addresses this idea, focusing on students at Stages 1 and 2, that is, on students at the third and fourth levels of fragmenting. So, the purpose of this chapter is to describe how these students think of fractions as parts in relation to wholes, to discuss goals for students at Stage 1, and to also discuss how to support students' progress toward these goals. Along the way we begin to introduce '$\frac{1}{m}$ of' fraction language to accompany 'out of' fraction language, with regard to unit fractions. For example, students who think about $\frac{1}{5}$ as 'one out of five equal parts' in the whole might also begin to consider $\frac{1}{5}$ as 'one-fifth' of the whole. In Chapter 6, we generalize that language to non-unit fractions.

Fractions as Parts Within Wholes

As we stated in Chapter 4, students at the first, second and third levels of fragmenting have not yet developed a way of thinking we call disembedding, which involves taking a part out of a whole while keeping the whole mentally intact.

Disembedding is an important component of conceiving of fractions as parts out of wholes. Prior to the development of disembedding, students view fractions as parts within wholes. So, students at Stage 1 are limited to developing a Parts Within Wholes Fraction Scheme (Hackenberg, 2013; Steffe and Olive, 2010). For these students, fractions are established as one part (or several parts) in a number of parts regardless of the relative size of the parts, and regardless of the inclusions a teacher may see. Viewing a partitioned whole in terms of part-to-part and part-to-whole relations is an important criterion for fractions knowledge (Piaget et al., 1960) that students at Stage 1 do not meet.

One example of the lack of disembedding can be seen in 7th-grade student Courtney's work to share a bar equally among four people (Hackenberg, 2013). Courtney marked the given bar into four parts. When asked about the sizes of the parts, she said that they were supposed to be equal. Then the teacher asked her to draw out the share for one person as a separate part. She did so, although her drawn part was about one-half of the width of the original bar (Figure 5.1).

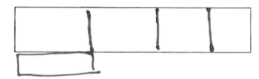

Figure 5.1 Courtney's equal sharing of a bar among four people and drawing out of one part

When the teacher asked for a fraction name for the part she had drawn out, Courtney said it was one-third (Hackenberg, 2013). She explained, 'You got three pieces left,' and she indicated that she took one piece away. Her response to this problem shows a lack of disembedding because when Courtney drew a part separate from the given bar, the four parts as a unit to which to compare that part seemed to become just three units. So, she seemed to mentally cut off a part from the bar and compare it to the three pieces that were left, but she did not disembed the part. To check this interpretation, the teacher stated that there were four parts in the bar. 'Is it fair to call this [the drawn out part] one out of four parts?' the teacher asked. Courtney paused. 'Or you're not sure?' said the teacher. 'I'm not sure,' said Courtney.

As another example, Olive and Vomvoridi (2006) have analyzed the case of Tim, who had constructed only a Parts Within Wholes Fraction Scheme in his 6th-grade year. At that time, one feature of Tim's concept of unit fractions was that both a unit fraction and the whole referred to the same partitioned image: one-sixth meant a whole partitioned into six equal parts, and six-sixths meant the same partitioned whole (Figure 5.2).

Figure 5.2   Tim's meanings for one-sixth *(left)* and six-sixths *(right)*

Source: Olive and Vomvoridi, 2006

In addition, Tim did not have a need for fractional parts to be equal in size. He demonstrated that the number of parts, regardless of size, yielded a unit fraction. For example, if there were four equal sevenths shown, Tim would call that entire amount 'one-fourth' (Olive and Vomvoridi, 2006: 25).

These ideas about fractions led Tim to add up parts regardless of size. For example, in adding one-half and one-fourth, Tim said the answer would be one-fifth because one-half was one part and one-fourth was four parts (Olive and Vomvoridi, 2006). Figure 5.3 gives an illustration of Tim's work.

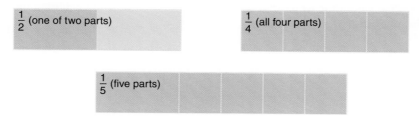

Figure 5.3   An illustration of Tim's idea that one-half plus one-fourth is one-fifth

Source: Olive and Vomvoridi, 2006

This example shows that constructing fractions as solely parts within wholes is severely limiting! We underscore that a student like Tim will not name one equal part of a whole as a unit fraction unless he or she has all the parts in his or her visual field. In other words, a part does not 'stand alone' for Tim and other students like him; it gains meaning only when all parts are present.

Since parts do not have a part-to-whole relation for Courtney, Tim and other students like them, justifications for having drawn a particular fraction rest upon numbers of parts: it is one-third because there are three parts left after I take one away; or, it is one-fifth because there are five parts drawn in all [regardless of size].

Fractions as Parts Out of Wholes

In contrast, once students have developed disembedding, they can learn to think of fractions as parts out of wholes, and they can develop a Parts Out of Wholes Fraction Scheme. This change in thinking signals a transition to Stage 2. For example, consider Grace, a 5th-grade student who had developed a Parts Out of Wholes Fraction Scheme and was asked to draw one-fifth of a rectangular cake. Grace partitioned the cake into five fairly equal parts, disembedded one of the parts from the whole and compared the part to the whole to establish a one-to-five comparison (Figure 5.4).

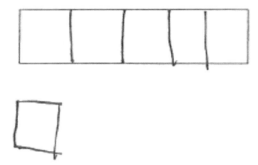

Figure 5.4   Grace's drawing of one-fifth of a given cake

Grace may also learn to do this with more parts – for example, she might learn to disembed three of the parts from the whole and compare them to the whole to establish a three-to-five comparison (Figure 5.5).

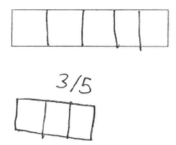

Figure 5.5   Three-fifths as three parts in comparison to five parts

Although this way of thinking and acting demonstrates a clear advance, conceiving of fractions as solely parts out of wholes is not sufficient for developing robust fractions knowledge. It is just a beginning. In particular, if Grace were asked to justify that a single part is $\frac{1}{5}$, she would likely rely on the idea that it is one of five equal parts. She would not have developed the idea that she could iterate that part five times to make the whole. We elaborate on this key point in Chapter 6, and there we will address working with Stage 2 students to progress beyond only part-to-whole comparisons.

Working with Students with Parts Within Wholes Fraction Schemes

Since conceiving of fractions as solely parts within wholes is so limiting, an important question is this: How can I as a teacher work productively with these students where they are, and how can I support these students to make progress? In our view, progress means that the students construct a disembedding operation, which is critical for constructing fractions knowledge in general. We note that students do not have to progress from Parts Within Wholes to Parts Out of Wholes Fraction Schemes: it is possible for a student to move from a Parts Within Wholes Fraction Scheme directly to a Measurement Scheme for Unit Fractions, a way of thinking that we address in Chapter 6.

To work toward that goal, engaging students in 'equal sharing problems', 'drawing unit fractions of quantities' and 'iterating unit fractions to make wholes' have been found to be especially helpful (Hackenberg, 2013; Olive and Vomvoridi, 2006). That is, the three problem types discussed in Chapter 4 are quite relevant here! This is not surprising, since in Chapter 4 we addressed students at the first three levels of fragmenting, and the third level corresponds to Stage 1 students. Here we elaborate on the third problem type so that we can highlight the mental action of iterating with Stage 1 students. We also discuss a fourth problem type, 'making comparisons of unit fractions'.

Some students at Stage 1 do not iterate when working with equal sharing problems or drawing unit fractions of quantities (Hackenberg, 2013). In that case, it can be helpful to provide students with a problem like this one, which can promote iteration.

'Finding the Equal Share Problem': Here is a cranberry cake (a rectangle) and several cut-out pieces that were made from identical cakes (Figure 5.6). (Next to the whole cake are the following parts: $\frac{1}{3}$, $\frac{1}{4}$, $\frac{1}{6}$ and $\frac{1}{8}$ of the cake. The parts are laid out on the table, unlabeled and not in order of size. It can be helpful to have very thin lines or arrows running parallel to the long edges of the cake so that students know which way to orient the parts.) The cake is to be shared equally among six people.

a)   Which one of these pieces would be one share?

b)   How do you know it's going to be an equal share?

Figure 5.6   Materials for 'Finding the Equal Share Problem' (from Hackenberg, 2013)

Students at Stage 1 may be able to select a part of appropriate size – perhaps even the correct part. However, when asked question (b), they may not move their selected part down the bar. For example, a 7th-grade student at Stage 1 named Laura selected the part that, from the teacher's perspective, was $\frac{1}{6}$. But when the teacher posed (b), Laura placed all of the four different given parts on the bar (Hackenberg, 2013).

The teacher then suggested to Laura that she use her selected part 'on the bar' to see whether it was an equal share (Hackenberg, 2013). Laura moved her part along the bar, marking each position with her finger before moving it to another position. As Laura moved the part along she left some space between each positioning of the part. So, she found that more than five parts and less than six parts fit. When asked to imagine that the parts fit exactly and that there were six parts, Laura named the part 'one-sixth'. This experience appeared to signal the beginning of iterating fractional parts for Laura. We note here that continuing to use the language 'one of six parts', as we discussed in Chapter 4, can be very useful for students even as they progress in their levels of fragmenting, because it refers to the actions for making the parts.

In addition to this kind of problem, Olive and Vomvoridi (2006) found that Tim (the 6th-grade student mentioned earlier) was helped by representing and iterating unit fractional quantities without the whole unit of 1 represented. Recall that, for Tim, 'six-sixths' and 'one-sixth' could refer to the same representation, a whole unit partitioned into six equal parts. That is, Tim did not seem to differentiate between one-sixth and six-sixths and saw them each as consisting of six parts, not necessarily equal. So, the teacher aimed for Tim to think of a unit fraction as the amount he needed to repeat some number of times to make the whole. To work on this goal, the teacher posed problems to Tim involving unit fractional quantities, such as this one:

> 'Ounces of Spices Problem' (adapted from Olive and Vomvoridi, 2006): You have two bags of spices. One bag contains one of ten equal parts ($\frac{1}{10}$) of an ounce of cinnamon and one bag contains one of five equal parts ($\frac{1}{5}$) of an ounce of nutmeg.
>
> a) How many bags of cinnamon do you need to make one ounce of cinnamon? Draw a picture to show.
>
> b) How many bags of nutmeg do you need to make one ounce of nutmeg? Draw a picture to show.
>
> c) Which weighs more, one bag of cinnamon or one bag of nutmeg? How do you know?
>
> d) How many bags of cinnamon, or how much of a bag, weigh(s) the same as one bag of nutmeg? How do you know?
>
> e) How many bags of nutmeg, or how much of a bag, weigh(s) the same as one bag of cinnamon? How do you know?

Tim's partner solved (a). The teacher engaged Tim in working on (b) (Olive and Vomvoridi, 2006). It took some time for Tim to determine that five bags of nutmeg made one ounce. At first he thought 10 bags of nutmeg made one ounce, and then he thought two bags did. The teacher discovered that he was thinking about how two bags of cinnamon weighed the same as one bag of nutmeg, part (d), because of the two-for-one relationship between the bags. In the process of thinking about this question, Tim realized that only five bags of nutmeg would make one ounce.

This problem brings up a fourth problem type that is useful for students at Stage 1: making comparisons of unit fractions. For example, consider this problem:

> Comparing Unit Fractions: At a fudge factory, the workers pour hot fudge into long rectangular trays and let it cool into slabs (Figure 5.7). Once the fudge has cooled, the workers slice a slab of fudge into some number of equal parts and put those pieces into display cases in the store. Two friends, Ming and Javier, came into the store and each selected a piece of fudge. Ming selected a piece that was $\frac{1}{5}$ ('one out of five parts') of a slab, and Javier selected a piece that was $\frac{1}{8}$ ('one out of eight parts') of a slab.
>
> a) Who selected a larger piece of fudge?
>
> b) How can you justify your response to (a)?
>
> c) Can you justify your response to (a) in another way?

Figure 5.7 A tray of fudge

This problem can allow students at Stage 1 to work on the inverse relationship between the number of parts created in a quantity and the size of those parts, i.e. an inverse relationship focused on partitioning. This kind of thinking may be a likely justification in part (b). However, by asking students for another way to justify in part (c), the teacher opens the possibility for students to think about iterating. That is, especially if students work on unit fraction comparisons following problems like the 'Ounces of Spices Problem', they may develop the idea that there is an inverse relationship between the number of times they need to iterate a part to make a whole unit and the size of that part.

Finally, students at Stage 1 are also very likely to benefit from many opportunities to reason with fraction bars (Olive and Vomvoridi, 2006). For example, Tim seemed to benefit from classroom instruction that used fraction bars to take fractions of fractions (e.g. find $\frac{1}{2}$ of $\frac{1}{3}$), the beginning of fraction multiplication. He also benefited from using fraction bars to make and visualize what we call commensurate fractions – two fractions that are the same size but are cut up into a different number of parts (e.g. $\frac{1}{4}$ and $\frac{5}{20}$). We address such opportunities in Chapters 10 and 11.

Assessment Task Groups

List of Assessment Task Groups

A5.1: Parts Within the Whole

A5.2: Parts Out of the Whole

A5.3: Finding the Equal Share

A5.4: Iterating a Part (adapted from Biddlecomb, 2002)

Task Group A5.1

Parts Within the Whole

Materials: Fraction strips and writing utensils; the fraction strips should all be the same length but also be marked into various numbers of equal parts between 3 and 9 (see the template in the Appendix).

What to do and say: Give each student a fraction strip and ask them to shade in two parts. Then ask them what fraction of the whole strip is shaded.

Notes

- This activity can be used to assess whether students can appropriately name fractions as parts within wholes. Although this conception of fractions is very limited, it can provide a starting point for talking about fractions in more sophisticated ways.
- Students should realize that, although they all shaded two parts, their fraction names are different because of the different numbers of equal parts in the whole.
- For students who do not name the fraction appropriately, it is okay to introduce naming conventions, but continue to use 'out of' language. For example, 'We write $\frac{2}{5}$ for this because we have two shaded parts that we can pull out of the five parts that make up the whole.' So, the written notation, $\frac{2}{5}$, means 'two out of five equal parts'.
- This activity can be repeated using other numbers of shaded parts. Because students will have different numbers of parts in their whole strip, they could be allowed to choose how many parts to shade and then to name the shaded fraction of the whole.

Task Group A5.2

Parts Out of the Whole

Materials: Fraction strips both marked and unmarked, writing utensils and scissors; the marked fraction strips should all be the same length but also be marked into various numbers of equal parts between 3 and 9.

What to do and say: Give students an unmarked and a marked fraction strip. Building off of A5.1, ask them to cut off two parts from the unmarked strip and name the fraction that amount is of the whole.

Notes

- This activity is intended to assess whether students maintain the part–whole relationship, even when the whole is cut.
- Let's say a student has used a strip marked into seven equal parts. Some students will name the two parts 'two-fifths' or 'two compared to five' because they no longer see the two parts that were cut off as parts of the whole. So, they compare the two parts that they cut off with the five parts remaining in the strip. In other words, they do not disembed. The instructional activities in the next section are intended to support students' development of a disembedding operation.

Task Group A5.3

Finding an Equal Share

Materials: A cardboard rectangle that represents a whole sub sandwich and a variety of shares made from this sandwich that are $\frac{1}{2}$ of the sandwich, $\frac{1}{3}$ of the sandwich, $\frac{1}{4}$ of the sandwich, $\frac{1}{5}$ of the sandwich, $\frac{1}{8}$ of the sandwich, etc. It can be helpful to have very thin lines or arrows running parallel to the longer edges of the sub sandwich so that students know which way to orient parts (see Figure 5.6).

What to do and say: Tell the students that the rectangle is a sub sandwich, and it is being shared among some number of people, say five. Which part shows the size of an equal share for one of five people? How do they know?

Note

- As an extension, ask about a number of equal shares that is not there among the cut-out parts; if it is not there, ask the students about how big it would be and how do they know. You can see whether they can determine that it would be smaller than a given part but bigger than another given part.

Task Group A5.4

Iterating a Part (adapted from Biddlecomb, 2002)

Materials: A cardboard rectangle that represents a whole sub sandwich and a variety of shares made from this sandwich that are $\frac{1}{2}$ of the sandwich, $\frac{1}{3}$ of the sandwich, $\frac{1}{4}$ of the sandwich, $\frac{1}{5}$ of the sandwich, $\frac{1}{8}$ of the sandwich, etc. It can be helpful to have thin horizontal lines running parallel to the longer edges of the sub sandwich so that students know which way to orient parts (see, e.g., Figure 5.6).

What to do and say: Select a part, such as $\frac{1}{3}$, and tell the students the part you have selected using fraction language ('one out of three equal parts' or you could try to use 'one-third'). Ask the students

(Continued)

(Continued)

how many of those parts are needed to make the whole sandwich. How did they know how many parts to use? Repeat with other parts. Then move to selecting a part and asking them how many are needed to make one of two equal parts (or 'one-half') of the whole sandwich. Again, ask how they knew how many parts to use.

Note

- With the question about making one of two equal parts of the sandwich, it is okay eventually to select parts that cannot be iterated a whole number of times to make one-half, such as the one-fifth part. Doing so has the potential to provoke a discussion about how to use one-fifth to make one-half. Most likely students at the first three levels of fragmenting will not have ideas about that. If a student does have an idea about 'use the one-fifth part two times and then one-half of that part because it looks like that will work', that can open the way to talking about taking unit fractions of unit fractions to determine $\frac{1}{2}$ of $\frac{1}{5}$, which we address in Chapter 10. It can also open the way to discussions about commensurate fractions (fractions the same size partitioned into different-sized parts). For example, two tenths are the same size as $\frac{1}{5}$, and five tenths are the same size as $\frac{1}{2}$. We address commensurate fractions in Chapter 11.

Instructional Activities

List of Instructional Activities

IA5.1:   Is It Fair?

IA5.2:   Tiny Creature Journeys

IA5.3:   Filling an Order

IA5.4:   Comparing Unit Fractions

IA5.5:   Finding Unit Fractions of Chocolate Bars

Activity IA5.1

Is It Fair?

Intended learning: Students will estimate the size of an equal share, thereby potentially introducing the idea of iterating (repeating) a part to 'measure' it along a bar.

Instructional mode: This activity can be played as a game with the whole class.

Materials: Fraction strips of the same size, scissors.

Description: The teacher gives each student (or pair of students) a fraction strip and then shows the class her strip. She asks the students to estimate the size of one fair share if the strip is shared among five people. Students cut off their estimate. Then individual students (or pairs of students) are invited to come to the front of the classroom to see whether their estimate is accurate.

Responses, Variations and Extensions

- The teacher can also **extend the range of numbers** by asking students to share the strip among more people, up to 10. Sharing among 10 people can support students' eventual understanding of **decimalizing** fractions.
- This activity can be extended by giving students a different-sized whole so that they can begin to understand that the size of the equal share will depend on the size of the whole. This understanding relates to **unitizing numbers**: the size of the unit fraction depends on the size of the original unit (the whole).

Activity IA5.2

Tiny Creature Journeys

Intended learning: Students will work on determining the number of unit fractions that make one whole and one-half of one whole, thereby potentially learning about repeating a unit fraction to 'measure' it along a bar.

Instructional mode: Students working individually with instruction from the teacher.

Materials: Paper, writing utensils, straws, scissors.

Description: Tell students that the ant travels one-third of a metre from the anthill to the green leafy bushes, and the ladybug travels one-sixth of a metre from her home to the green leafy bushes. Give students a straw that is to represent one-third of a metre and call it the 'ant journey'. How many ant journeys does it take to make a whole metre? How do you know? Ask students to draw a picture to show. Repeat for the 'ladybug journey', but this time don't give a straw. Ask which is longer, an ant journey or a ladybug journey. Ask how many ant journeys, or parts of an ant journey, make a ladybug journey. Ask how many ladybug journeys, or parts of a ladybug journey, make an ant journey. After this, ask students to make a straw that represents a ladybug journey accurately, in relation to the straw that represents the ant journey.

Responses, Variations and Extensions

- This is the 'Ounces of Spices Problem' recast in terms of length. The reason to use lengths is that they are a more intuitive quantity for students than weight.
- However, one advantage of weight is that small differences are hard to see (it is hard to visually tell that a bag contains $\frac{1}{5}$ of an ounce v. $\frac{1}{10}$ of an ounce of something). In contrast, with lengths relative size is more easily seen, within a certain range of values and units. So, this is one reason not to give a straw for the ladybug journey. Both the use of weights and withholding the straw can be means of **distancing the instructional setting**.
- If students struggle to respond to the questions for the ladybug journey, you can give another representation of a length – say a popsicle stick, that is not exactly one-half of the straw length. If it is exactly one-half, then students will be tempted to assess visually without thinking about the relationships to the whole metre.

(Continued)

(Continued)

- Students at Stage 1 may find the relationships between the two journeys hard to identify; they also may find making a straw to represent the ladybug journey difficult. If so, don't push that on them – use the information as formative assessment of where students are, instead of as a sign to push certain ways of thinking.
- Drawing pictures will provide opportunities for students to engage in notating. For example, students might label their drawings with informal notations like the following: '2 ladybugs = 1 ant'. Those notations relate to later formalizing wherein $2(\frac{1}{6}) = \frac{1}{3}$.

Activity IA5.3

Filling an Order

Intended learning: Students will work on determining the number of unit fractions that make one whole and one-half of one whole in a weight context.

Instructional mode: Students working in pairs with instruction from the teacher.

Materials: Paper, writing utensils, a picture of a two-pan balance scale or an actual two-pan balance.

Description: Ask students if they have ever been with their parents in a grocery store where items like nuts, dried fruit, rice or beans are sold in bulk. Customers take a bag and open a bin to scoop out the items, or pour them from a spigot, and then they weigh their bag on a scale to determine the price. The students are all workers in a grocery store where the scale is malfunctioning! The store manager has asked them all to create unit fractions of 1 pound of every bulk item. The teacher assigns a different bulk item to each student and with a unit fraction. For example, in a pair one student might be 'pinto beans, $\frac{1}{5}$ pound' and another student might be 'brown rice, $\frac{1}{10}$ pound'. The customers coming into the store that morning all want to buy a whole pound of their bulk items. Ask the student to draw a picture to show how many bags of their item are needed to make a whole pound and to explain their solutions to their partners. Ask pairs of students to compare the number of bags needed of each of their items and explain why they are different amounts.

Responses, Variations and Extensions

- This is another version of the 'Ounces of Spices Problem', using weights. As noted in IA5.2, the use of weights can be a means of distancing the instructional setting.
- Teachers can ask similar questions to those used in the 'Ounces of Spices Problem', as long as the partners have fractions that are whole number multiples of each other.

Activity IA5.4

Comparing Unit Fractions

Intended learning: Students will work on size relationships for unit fractions.

Instructional mode: Students working in pairs with instruction from the teacher.

Materials: Fraction strips (all the same length), writing utensils, scissors, small white boards and markers (one white board for each student).

Description: Give each student in a pair a unit fraction, say $\frac{1}{4}$ to one student and $\frac{1}{6}$ to another student. The students are to tell each other their fractions, each predict which fraction is bigger, and record their prediction on the white board. Then they are each to make their unit fraction with the fraction strips. For example, a student might fold the strip into four equal parts and then cut off one part. Ask the students to compare their strips and discuss what they have made. Then ask students to explain what they have found out, to give reasons for what they have found out and to reflect on their initial predictions.

Responses, Variations and Extensions

- This activity can be made into a game, with students earning points for accurate predictions but also for good explanations of why they would change their initial prediction.
- Students should begin to recognize the reciprocal relationship between the number of equal parts indicated by the unit fraction and the number of segments in each bar. As such, teachers can use this activity to support **generalizing**, wherein students begin to understand why the unit fraction with the larger denominator will be the smaller fraction.

Activity IA5.5

Finding Unit Fractions of Chocolate Bars

Intended learning: Students will join their whole number knowledge and partitioning knowledge to make unit fractions.

Instructional mode: Students working individually with instruction from the teacher.

Materials: Paper with a set of segmented equal chocolate bars on it: a unit, 3-bar, 4-bar, 5-bar, 6-bar, 8-bar, 10-bar and 12-bar. The bars should be marked into equal parts.

Description: The teacher tells students that he will name some unit fractions of different chocolate bars. Students are to find the bar that represents that fraction on their paper, or they are to draw it if it is not there. For example, the teacher might ask students to find 'one of two equal parts of the 6-bar, or one-half of the 6-bar'. Students locate that bar or draw it, and justify how they know what they found or drew is one of two equal parts of the 6-bar. Then the class can discuss student responses. The teacher can then ask students to find or draw 'one of three equal parts of the 6-bar, or one-third of the 6-bar', etc.

Responses, Variations and Extensions

- This task builds on the idea of connected number (see IA4.4) by asking students to make unit fractions of whole number lengths.
- As students make progress the teacher can use chocolate bars with more parts, **extending the range of numbers**.

6

Teaching Students at Stage 2: Measuring with Unit Fractions

The Parts Out of Wholes Fraction Scheme provides students with a foundation for further development of fractions knowledge. The dilemma for many students is that they never develop more sophisticated knowledge and so they continue to rely on part–whole concepts throughout their schooling experiences, which limits their opportunities to learn from those experiences (Olive and Vomvoridi, 2006). Indications of this are present among adults, too, who are uncomfortable working with improper fractions, such as $\frac{7}{5}$, and reflexively convert them to mixed numbers (1 and $\frac{2}{5}$) or decimals (1.4). After all, 7 out of 5 makes no sense!

In the United States, textbooks emphasize part–whole concepts to the exclusion of other concepts and to the detriment of students' education (Watanabe, 2007). This emphasis stands in contrast to textbooks in East Asian countries (Li et al., 2009; Watanabe, 2007; Yang et al., 2010) and the Common Core State Standards (2010) in the United States.

Common Core State Standards for Mathematics (CCSSM) call on educators in the United States to begin supporting students' development of measurement concepts for fractions as early as 3rd grade, including the placement of fractions on a number line:

> CCSS.MATH.CONTENT.3.NF.A.2 Understand a fraction as a number on the number line; represent fractions on a number line diagram.

However, a number line only begins to make sense in reference to an explicit unit of measure. For whole numbers on the number line, the unit of measure is 1. Assessment and instructional activities in Chapters 4 and 5 can support students in developing the idea of a connected number as a precursor to a number line (see A4.4 and IA4.4, and IA5.5). Making connected numbers (i.e. a sequence of continuous wholes by iterating a continuous unit so many times) can be the beginning of treating 1 as a unit of measure.

To place fractions on the number line, we need new units of measure: unit fractions, like $\frac{1}{5}$. In this sense, we can think of unit fractions as fractional units. They are sizes or quantities relative to the whole, and they can be used to measure off non-unit fractions. This, too, aligns with the CCSSM, which demand that 'students view fractions in general as being built out of unit fractions'.

This chapter focuses on unit fractions as fractional units, or units of measure. Once students develop such a concept for unit fractions, they can re-conceptualize non-unit fractions as well, through the process of measuring them with unit fractions. These new conceptions have implications for how students can think about fractions. When thinking about a non-unit fraction, such as 3 out of 5 equal parts in the whole, students who have developed fractional units can begin to think about it as three-fifths of the whole – a size relation, rather than a part–whole relation.

Unit Fractions and the Whole

Students can use their Parts Out of Wholes Fraction Schemes to recognize a unit fraction, such as $\frac{1}{5}$, as one part out of five equal parts in the whole. Furthermore, they can rely on mental actions of disembedding and iterating (addressed in Chapter 5) to pull $\frac{1}{5}$ out of the whole and reproduce the whole by making five connected copies of that $\frac{1}{5}$ part. As illustrated in Figure 6.1, there is a kind of reversibility built into this activity: students produce $\frac{1}{5}$ from the whole by partitioning the whole into five parts and pulling out one of those parts (top to bottom in Figure 6.1); and, conversely, they can produce the whole from $\frac{1}{5}$ by taking that part and iterating it five times (bottom to top in Figure 6.1). These are the kinds of experiences students need in order to conceptualize unit fractions as fractional units, i.e. as units of measure.

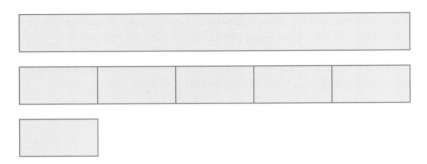

Figure 6.1   Partitioning wholes and iterating unit fractions

Activities like this help establish a multiplicative relationship between the size of the unit fractional part and the whole. In this case, the whole is five times as big as the unit fractional part. There is a 5-to-1 relationship that requires coordinating the two levels of units – the size of the unit fractional part and the whole. We refer to this way of working with unit fractions as the *Measurement Scheme for Unit Fractions* (note that, elsewhere, this scheme is referred to as the Partitive Unit Fraction Scheme; see Steffe, 2002).

The Measurement Scheme for Unit Fractions

As a reversible scheme, the Measurement Scheme for Unit Fractions works in two directions. Students who have developed the scheme can use it in cases where they want to make a unit fraction from a given whole. They can also use it to make the whole from a given unit fraction of that whole. Beyond the Parts Out of Wholes Fraction Scheme, which involves the mental actions of partitioning and disembedding, the Measurement Scheme for Unit Fractions also involves iterating. It is closely related to equi-partitioning, mentioned in Chapters 2 and 3.

The Role of the Whole

Fractions have meaning only in relation to a specific whole. With measurement schemes, that relation is a multiplicative size relation. In the absence of such a relation, students might think that $\frac{1}{7}$ is bigger than $\frac{1}{6}$. For example, consider Isaac – a 6th-grade student in the United States who had constructed a Parts Out of Wholes Fraction Scheme but had not constructed any measurement schemes. Thus, he could produce proper fractions of a whole, but when asked to compare the sizes of fractions – even unit fractions – he responded in surprising ways.

When asked to compare $\frac{1}{7}$ and $\frac{1}{6}$, Isaac claimed that $\frac{1}{7}$ was bigger. When asked to justify his answer, he drew the picture shown in Figure 6.2. Isaac did not appear to have a referent whole in mind when he interpreted $\frac{1}{6}$ or $\frac{1}{7}$. As a result, he drew each one-sixth part the same size as each one-seventh part. His conception of these fractions seemed to be based on the number of pieces – not their size or their relation to the whole. He showed that $\frac{1}{7}$ was bigger than $\frac{1}{6}$ because seven pieces made a longer length than six pieces. In fact, $\frac{1}{7}$ *could* be bigger

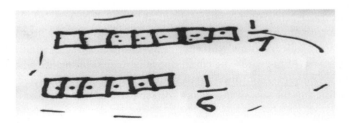

Figure 6.2 Isaac's comparison of $\frac{1}{6}$ and $\frac{1}{7}$

Source: Norton and Boyce, 2013

than $\frac{1}{6}$ if $\frac{1}{7}$ came from a longer whole and $\frac{1}{6}$ came from a shorter whole. Implicit in the comparison of fractions is that they come from the same-size whole.

Another example from Isaac demonstrates that simply drawing the whole does not resolve the issue. The teacher asked Isaac which was bigger, one-eighth of a pizza or one-fourth of a pizza. Isaac drew the picture shown in Figure 6.3, partitioning the whole into eight parts and shading four of these parts. This picture supported Isaac's notion that $\frac{1}{8}$ was bigger (eight parts is more than four parts). Moreover, Isaac recognized the four shaded parts as one-half of the whole pizza and, based on this recognition, claimed that sometimes $\frac{1}{4}$ and $\frac{1}{2}$ are the same! The only way to resolve such issues is to hold in mind the unit fraction and the whole at the same time.

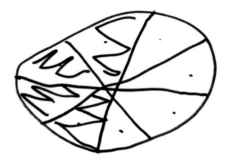

Figure 6.3 Isaac's comparison of $\frac{1}{4}$ and $\frac{1}{8}$

Source: Norton and Boyce, 2013

Reciprocal Relationship Between the Unit Fraction and the Whole

Through experiences like the instructional activities that appear later in this chapter, Isaac did eventually begin coordinating unit fractions and wholes. At the same time, he seemed to notice the reciprocal relationship between the number of parts in a whole and the size of each part: 'You're going to have to make the pieces smaller [to fit more into the whole].' Isaac had been associating unit fractions with the number of them that made up the whole, but without fixing the size of the whole he could not reliably determine the size of the unit fractional pieces. Once he began coordinating the number of pieces with a fixed whole, he could also begin coordinating the size of the pieces with the size of the whole and comparing unit fractional sizes.

Understanding the reciprocal relationship between the number of pieces in the whole and the size of each piece relies upon coordinating the two units: the unit fraction and the whole. In turn, this understanding supports the concept of unit fractions as fractional units that can be used to measure other fractions. Gould (2012) referred to this concept as 'quantity fractions' because it gives meaning to fractions as quantities (sizes or measures), whereas part–whole concepts support only fractions as parts.

Measuring Non-Unit Fractions

Once unit fractions become fractional units for a student, they can begin using unit fractions to measure non-unit fractions. For example, $\frac{3}{5}$ is not merely three out of five equal pieces disembedded from the whole; it is also the quantity formed by taking three measures of a $\frac{1}{5}$ unit, as shown in Figure 6.4. However, as we will see in the next two chapters, the situation becomes more complicated when students iterate the unit fraction beyond the whole to make an improper fraction, like $\frac{7}{5}$.

Figure 6.4 $\frac{3}{5}$ as three $\frac{1}{5}$ s

Comparing Fractions as Quantities

As multiples of a unit fraction, proper fractions can begin to take on quantitative meaning. Students can start comparing the sizes of proper fractions to one another. When the fractional unit is the same, the comparison is straightforward (e.g. $\frac{4}{5}$ is bigger than $\frac{3}{5}$ because it is four measures of the unit versus three measures of that same unit). Students who conceptualize fractional units might also compare proper fractions when the number of measures is the same and the fractional units are different (e.g. $\frac{3}{4}$ is bigger than $\frac{3}{5}$ because the $\frac{1}{4}$ unit is bigger than the $\frac{1}{5}$ unit). Of course, students can also learn to make these comparisons using rules (e.g. if the numerators are the same, the bigger fraction is the one with the smaller denominator), but now students can develop meaning for these rules. They could even rely on that meaning to develop the rules themselves.

Without meaning, students will often generalize rules in inappropriate ways. For example, in comparing $\frac{3}{5}$ to $\frac{4}{7}$, a student might think $\frac{4}{7}$ is bigger because it has the larger numerator. Conversely, they might think that $\frac{5}{6}$ is smaller than $\frac{3}{5}$ because it has the larger denominator. Of course, students can learn new rules for making such comparisons. For example, cross-multiplying tells us that $\frac{5}{6}$ is bigger than $\frac{3}{5}$ because 5 × 5 is bigger than 6 × 3 (see Figure 6.5). But doing mathematics is not memorizing rules. It involves meaningfully applying concepts to solve problems and generate new ideas.

Figure 6.5   Cross multiplying

Students can rely on measurement schemes for unit fractions to generate many new ideas regarding fraction comparison. Consider the previous example of $\frac{5}{6}$ and $\frac{3}{5}$. Understanding that these fractions are sizes related to the same whole, students might consider their complements: $\frac{5}{6}$ needs $\frac{1}{6}$ more to complete the whole; $\frac{3}{5}$ needs $\frac{2}{5}$ more to complete the whole. Because $\frac{1}{6}$ is smaller than $\frac{2}{5}$, $\frac{5}{6}$ must be closer to the whole than $\frac{3}{5}$ and, therefore, $\frac{5}{6}$ must be bigger than $\frac{3}{5}$ (see Figure 6.6).

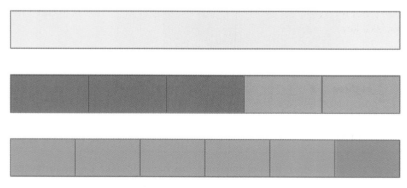

Figure 6.6   Comparing $\frac{5}{6}$ and $\frac{3}{5}$

Commensurate Fractions

Following Steffe (2004), we refer to equally sized fractions of a whole as 'commensurate fractions', rather than equivalent fractions, because equivalence implies something more. Whereas 'commensurate' means that two fractions have the same measure, 'equivalent' is a formal term that refers to an equivalence class.

When students begin thinking of fractions as sizes, measures, or quantities, understanding commensurate fractions becomes an important consideration. Whereas part–whole concepts for $\frac{3}{12}$ and $\frac{1}{4}$ might make these fractions seem different, as measures, they are the same: $\frac{3}{12}$ and $\frac{1}{4}$ are just different names for the same fraction. Understanding this relationship relies upon seeing three twelfths within each one-fourth of the whole – a 3-to-1 relationship between the fractional unit of $\frac{1}{12}$ and the fractional unit of $\frac{1}{4}$.

Students who learn to coordinate fractional units can use that way of working to make sense of fraction addition and fraction multiplication, as well as comparing fractions. However, because students must also coordinate these units with the whole, the coordination involves three levels of units. Students operating at Stage 2 can coordinate a third level of units, but they must do so through some kind of activity. In Chapter 11, we will consider different ways that students at Stages 2 and 3 work with commensurate fractions.

Assessing Measurement Conceptions of Unit Fractions

We have argued that students at Stage 2 can make significant progress in understanding fractions, including fraction comparison. Here we provide assessment tasks to distinguish conceptual development within Stage 2. Then we describe instructional activities to support that development.

Tasks for assessing measurement conceptions concentrate on fractions as sizes relative to the whole. Beyond disembedding unit fractions from partitioned wholes, these tasks elicit student actions of iterating unit fractions. For example, given a one-fifth piece, students should understand that they can reproduce the whole by iterating that piece five times. While implementing assessment tasks, key considerations for the teacher include the following:

- Do students understand fractions as only parts out of wholes, or can they also understand fractions as quantities?
- Can students make unit fractions from the whole and, conversely, make the whole from a given unit fraction?
- Do students understand the reciprocal relationship between the number of parts in a whole and the size of each part?
- Can students use unit fractions as fractional units with which to measure non-unit fractions?
- Can students generalize their measurement conceptions from unit fractions to non-unit fractions?

Instruction on Fractions as Measures

Understanding unit fractions as measures relies upon the reversibility of partitioning and iterating. In particular, students should be able to produce unit fractions by partitioning the whole, and they should be able to reproduce the whole from a unit fraction by iterating it. By performing these activities for various fractions, students begin to understand how unit fractions relate to the whole as relative sizes – including the reciprocal relationship between the number of equal parts in a whole and the size of each part.

Once teachers understand their students' ways of working with unit fractions, they can intentionally help students develop more robust conceptions of unit fractions as measures. Teachers can provoke this development by using instructional tasks that elicit iterating unit fractions in new ways and for new purposes, including the production of non-unit fractions from unit fractions. For example, students can produce $\frac{3}{5}$ from $\frac{1}{5}$ by iterating the unit fraction, $\frac{1}{5}$, three times. Such tasks require students to coordinate a third level of units in activity.

Understanding unit fractions as measures opens possibilities for meaningfully operating on fractions. However, beyond the two-level coordination between unit fractions and wholes, arithmetic operations that combine two fractions, such as adding and multiplying, require the coordination of a third level of units. Because Stage 2 students take only two levels of units as given, they must coordinate this third level through some kind of activity. Thus, instructional tasks for supporting Stage 2 students' understanding for adding and multiplying fractions often need to include some kind of manipulative (setting). These issues are discussed in Chapters 10 and 11.

Assessment Task Groups

List of Assessment Task Groups

A6.1:   Naming Unit Fractions

A6.2:   Producing Unit Fractions from the Whole

A6.3:   Producing the Whole from Unit Fractions

A6.4:   Producing Non-Unit Fractions from Unit Fractions

Task Group A6.1

Naming Unit Fractions

Materials: Fraction strips (long, thin rectangular strips of paper; see Appendix).

What to do and say: Ask the student to fold a strip into two equal parts. Then ask them to describe the size of one of the parts compared to the whole. Ask them to do the same thing with three, four, five and six equal parts.

Notes

- The purpose of this task is to assess whether the student will persist in creating equal parts and will appropriately name one of those parts based on its size relative to the whole.
- Student responses indicate a measurement concept for unit fractions if they name each part based on how many times it fits into the whole (e.g. 'the part is one-fourth because it could fit into the whole four times').

Task Group A6.2

Producing Unit Fractions from the Whole

Materials: A variety of rectangles made from cardboard – multiple copies of the same size is helpful; writing utensil for making marks; scissors for cutting bars; a pair of dice.

What to do and say: Working individually, students will select two copies of a rectangle that represents a piece of lumber (a board). The teacher rolls the dice to determine the unit fraction of the board that students should cut off. For example, if the dice total is 7, students will be asked to cut off one-seventh of the whole board. So everyone in the class is making the same unit fraction, but not necessarily the same-sized piece. Students are to make *one* cut and show that piece to the teacher, along with the other copy of the whole board.

Notes

- This task is like Instructional Activity IA4.2, from Chapter 4. Here, we use it as an assessment task, to determine students' progress in developing a measurement scheme for unit fractions. Also, by using a pair of dice, there will be a larger range of unit fractions.
- Even if their initial response is inaccurate, students at Stage 2 should be able to use their iterating operation to determine whether the piece they cut off is too big or too small.
- More accurate cuts may indicate that the measurement scheme for unit fractions is well established for the student.

Task Group A6.3

Producing the Whole from Unit Fractions

Materials: Fraction rods (also known as 'Cuisenaire rods'; see Appendix for template).

What to do and say: Show the student a red rod (two units long) and tell them that it is one-third of the whole. Ask them to identify which rod is the whole. Similar questions include the following: 'If the red rod is one-fifth of the whole, find the whole'; 'If the white rod is one-seventh of the whole, find the whole'.

(Continued)

(Continued)

Notes

- Students who understand unit fractions as fractional units should be able to find the appropriate whole by finding a rod that is n times as long as the $\frac{1}{n}$ rod. In the given examples, the whole rods are the dark green rod, the orange rod and the black rod, respectively.
- If students line up $n\frac{1}{n}$ rods to make the length of the whole, ask them to try to find the whole using only one of the $\frac{1}{n}$ rods. This may provoke them to focus on iterating a length rather than producing n parts.

Task Group A6.4

Producing Non-Unit Fractions from Unit Fractions

Materials: Fraction rods.

What to do and say: Show the student a light green rod (three units long) and tell them that it is $\frac{1}{3}$. Ask the student to identify which rod is $\frac{2}{3}$. Similar questions include the following: 'If the red rod is $\frac{1}{5}$, which rod is $\frac{3}{5}$?'; 'If the white rod is $\frac{1}{8}$, which rod is $\frac{3}{8}$?'

Notes

- These tasks are similar to those in Task Group A6.3, except students need to iterate the given unit fraction to make a non-unit fraction, rather than iterating it to make the whole.
- Students with measurement schemes should be able to produce the non-unit fractions using just a single copy of the unit fraction.

Instructional Activities

List of Instructional Activities

IA6.1:   Comparing Unit Fractions

IA6.2:   Guess My Fraction (Fractions as Measures)

IA6.3:   Left, Right, or Just Right (Placing Fractions on a Number Line)

IA6.4:   Different but the Same (Commensurate Fractions)

Activity IA6.1

Comparing Unit Fractions

Intended learning: Students will learn to compare the sizes of unit fractions.

Instructional mode: Game that can be played with the whole class.

Materials: Fraction strips (all the same length; see Appendix for template), pencil, scissors, a pair of dice.

Description: Students will make unit fractions using fraction strips and compare the size of their fractions to those of other students. For each round, students will roll a pair of dice to determine a denominator for their fraction. Students will then use pencils and scissors to make their fraction. For example, if a student rolled a 9, he might mark the strip into nine equal parts and cut off one of them. The teacher will encourage students to name their fractions and compare their sizes to those of the students around them. The teacher can ask them questions like the following: 'Who thinks they have the biggest fraction, and why?' 'Who has the smallest?' 'Who has $\frac{1}{7}$; who has $\frac{1}{10}$; and how can we compare them?' 'Does anyone have a conjecture about which fractions will be bigger and which fractions will be smaller?'

Responses, Variations and Extensions

- This activity resembles IA5.4, in which students should have begun to recognize the reciprocal relationship between the size of the denominator and the size of the unit fraction. We can extend the activity further – for Stage 2 students – in the following ways.
- Teachers can **extend the range of numbers** by telling students they are to make two of the unit fraction. For example, if a student rolled a 7, they would make $\frac{2}{7}$, rather than $\frac{1}{7}$.
- This game can extend further by rolling the dice to determine both numerators and denominators for the fractions. Each student would roll the dice once; the smaller number would be the numerator, and the larger number would be the denominator.
- These extensions can also help students to **structure fractions as numbers**. For example, $\frac{2}{5}$ can be structured as two of $\frac{1}{5}$ and not just two out of five.

Activity IA6.2

Guess My Fraction (Fractions as Measures)

Intended learning: Students will learn to coordinate partitioning and iterating actions to make fractions of specified sizes.

Instructional mode: Game that can be demonstrated with the whole class and then played in small groups.

Materials: Fraction rods (also known as 'Cuisenaire rods'; see Appendix for template).

Description: One student secretly chooses a rod and hides it behind her back. Then she gives the other students a hint about it using another rod. The other students have to guess the size of the hidden rod. For example, the first student could choose the brown rod, hide it behind her back, and then show the other students a red rod saying, 'this rod is one-fourth as big as the rod I chose'.

Responses, Variations and Extensions

- The kinds of hints students give indicate their sophistication in thinking about fractions as sizes. The game becomes more or less challenging for other students depending on these hints.
- There are two educational apps related to this game and available for free on iTunes. Both are called *CandyFactory*. The older version works on iPhones, iPod touches and iPads; the newer

(Continued)

(Continued)

> version only works on iPads. Information on how to use the apps can be found at http://ltrg. centers.vt.edu and in Norton et al. (2014).

- This activity can be used to support standards like Common Core State Standard 3.NF.1, which demands that students understand a non-unit fraction, $\frac{a}{b}$, as 'the quantity formed by a parts of size $\frac{1}{b}$'. It also aligns with a similar standard for grade 4, 4.NF.4, which calls for students to understand a/b as a *multiple* of $\frac{1}{b}$.

Activity IA6.3

Left, Right, or Just Right (Placing Fractions on a Number Line)

Intended learning: Students will place proper fractions on a number line from 0 to 1.

Instructional mode: Short, whole-class activity that the teacher should conduct periodically.

Materials: White board and markers.

Description: The teacher draws a line segment on the board (varying the length each time) and labels 0 and 1. Students pick a fraction between 0 and 1, and one at a time, they label its position on the number line. For each label, the class votes on whether this is too far left, too far right, or just right: 'Left', 'Right', or 'Just Right'. Students are given a chance to explain their votes. Once a fraction is labeled, that fraction cannot be used again by other students.

Responses, Variations and Extensions

- Teachers should let lower-performing students go first so that they might label simpler values, like $\frac{1}{2}$ or $\frac{1}{4}$.
- For higher-performing students, the teacher can challenge them by asking questions like 'Who can think of a fraction really close to 1?'
- This goal aligns with standards that call for students to treat fractions as numbers on a number line, like Common Core State Standard 3.NF.2.
- Placing fractions on a number line can further support students in structuring numbers.

Activity IA6.4

Different but the Same (Commensurate Fractions)

Intended learning: Students will learn how to produce various fractions that are commensurate with a given unit fraction.

Instructional mode: Students working individually with instruction from the teacher.

Materials: Fraction strips (all the same length), coloured markers, scissors, tape.

Description: The teacher will specify a unit fraction. Students will make and cut out the fraction. The students should keep that cut-out fraction along with another sheet fraction strip (the whole). The teacher will ask the students to name as many commensurate (same-sized) fractions as they can, by making partitions of the fraction and the whole using a marker. Students will write down the commensurate fraction with that marker and then make another commensurate fraction using a different-coloured marker.

Responses, Variations and Extensions

- This goal aligns with Common Core State Standards 3.NF.3 and 4.NF.1, which demand that students identify equivalent fractions by using visual models.
- Tasks in Chapter 11 will extend commensurate fractions beyond those based on a unit fraction.

7

Teaching Students at Stage 2: Reversible Reasoning

As discussed in Chapter 6, the Measurement Scheme for Unit Fractions enables students to begin conceptualizing fractions as sizes, where unit fractions become units of measure. Reversibility is built into that scheme in the sense that students can work from the unit fraction to the whole or from the whole to the unit fraction. For example, students can rely on the Measurement Scheme for Unit Fractions to produce one-fifth of a given whole by partitioning the whole into five equal pieces, and to reproduce the whole from a given one-fifth piece by iterating it five times (see Figure 7.1).

Mental actions of partitioning and iterating form the basis for reversibility in the Measurement Scheme for Unit Fractions. However, at first students have to perform these actions sequentially. In the example provided above, the whole can be partitioned into five parts, and then one of those parts can be iterated five times to reproduce the whole. For students to develop complete reversibility, they need to anticipate these actions as inverses and think about them simultaneously, as two sides of the same coin. This simultaneity generates a single mental action called *splitting* – the operation that allows students to reverse their reasoning with fractions (Hackenberg, 2010; Steffe, 2002).

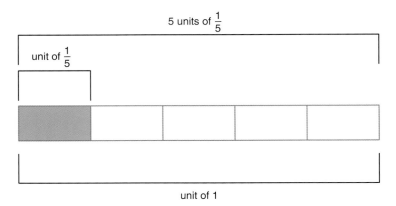

Figure 7.1   Reversibility and the Measurement Scheme for Unit Fractions

Source: Boyce and Norton, 2016

The Splitting Operation

Splitting combines the mental actions (operations) of partitioning and iterating into a single operation (Norton, 2008; Steffe, 2002; Wilkins and Norton, 2011). Students who can split do not need to iterate or partition in order to see how one action undoes the other. For this reason, they are able to solve tasks like the ones shown in Figure 7.2. Although the tasks are iterative in nature – referring to something that is *n* times as big – the solution requires students to partition, anticipating that partitioning will undo the hypothetical action of iterating. In this sense, partitioning and iterating become a single operation that combines the two more basic operations.

The stick shown below is three times as long as another stick. Draw the other stick.

The amount of pizza shown below is six times as big as your slice. Draw your slice.

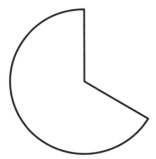

Figure 7.2   Splitting tasks

Splitting Task Responses

Students who do not split will often interpret splitting tasks as iterating tasks. Rather than reversing the iteration by partitioning the given part, they might iterate the given part. For example, consider the student response shown in Figure 7.3. The student appeared to iterate the given stick three more times and then to correct their solution so that there were a total of three iterations (including the given stick). The student did not interpret the given stick as a result of iterating some other stick three times. Reversing the iteration in this way – and, thus, partitioning the given stick into three equal parts to produce the other stick – requires the mental action of splitting.

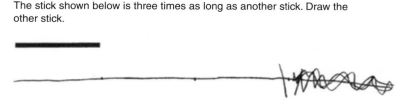

The stick shown below is three times as long as another stick. Draw the other stick.

Figure 7.3   Splitting task responses (from Wilkins and Norton, 2011)

Reversible Reasoning with Proper Fractions

Recall that the Measurement Scheme for Unit Fractions enables students to build non-unit proper fractions from unit fractions by iterating the latter as units of measure. For example, $\frac{3}{5}$ is three iterations of the unit $\frac{1}{5}$, which is the result of partitioning the whole into five equal parts. With splitting, students can anticipate how to reverse the operations of iterating and partitioning in order to produce the whole from a given proper fraction (Hackenberg, 2010). To reproduce the whole from $\frac{3}{5}$ involves partitioning $\frac{3}{5}$ into three equal parts and then iterating one of those parts five times (see Figure 7.4).

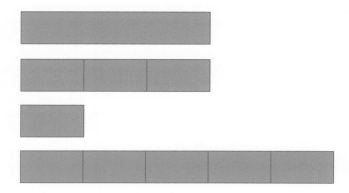

Figure 7.4   Reversing proper fractions

This kind of reversible reasoning requires students to coordinate three levels of units: the proper fraction, the unit fraction and the whole (see Figure 7.5). However, because the proper fraction is contained in the whole, students can coordinate these three levels of units in activity, by working within the whole (Hackenberg, 2007). Thus, students at Stage 2 can begin to reverse proper fractions once they can split.

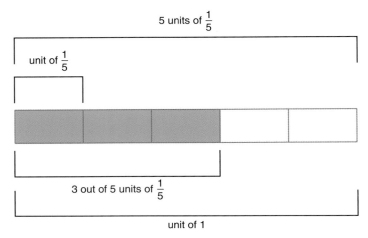

Figure 7.5   Three levels of units involved in reversing proper fractions

Source: Boyce and Norton, 2016

Coordinating Three Levels of Units in Activity

Students at Stage 2 can take two levels of units as given and build up a third level of units through activity (see Chapters 2 and 3). In reversing fractions, students take the proper fraction as a two-level unit; for example, $\frac{3}{7}$ is a unit of three sevenths. Having obtained one-seventh through splitting, a student can then build up the third level of units – the whole – through the activity of iterating that one-seventh piece seven times, or by adding on four more sevenths to the three-sevenths piece.

We note that reversing improper fractions requires more from the student because the fraction exceeds the whole (Hackenberg, 2007). For instance, if a student wanted to produce the whole from $\frac{9}{7}$, she would need to keep track of the whole within the fraction and keep those two units distinct. Otherwise, she might partition the improper fraction into nine parts and think she has produced ninths of the whole, or she might erroneously partition the fraction into seven parts. In other words, to consistently and meaningfully solve such tasks, she needs to simultaneously coordinate three levels of units – the improper fraction, the whole and the unit fraction. This is the purview of Stage 3 students and the focus of Chapter 8.

Assessing Reversible Reasoning

As discussed in Chapter 6, students at Stage 2 can coordinate mental actions of partitioning and iterating to begin conceptualizing fractions as measures or sizes. Furthermore, these students are in a position to construct a splitting operation, which enables them to reverse their reasoning about fractions as sizes; they can produce a proper fractional size relative to a given whole, and they can reproduce the whole from a given proper fraction. However, this way of operating does not extend to improper fractions until students can simultaneously coordinate three levels of units (Stage 3).

Because splitting is the key operation for reversible reasoning, it is important that teachers be able to assess whether students have constructed this operation. Appropriate tasks for assessment include those illustrated in Figure 7.2. While implementing such tasks, key considerations for the teacher include the following:

- Do students anticipate the inverse relationship between iterating and partitioning, or do they have to coordinate these operations through activity?
- Are students able to coordinate three levels in activity? Specifically, can they relate a proper fraction to its unit fraction and the whole?
- Can students work from the fraction to the whole, as well as the whole to the fraction?

Instruction on Reversible Reasoning

The construction of a splitting operation involves a reorganization of students' existing operations of partitioning and iterating. This reorganization is an internal and subconscious process that cannot be directly taught. Rather, research indicates that students' ability and experiences in coordinating partitioning and iterating activities (the focus of Chapter 6) lead to this reorganization over time (Norton and Wilkins, 2013). Thus, there is no instructional intervention for splitting. However, once a teacher has assessed that a student can split, numerous instructional activities become available for supporting that student's development of reversible reasoning.

As we have discussed, students who can split can begin solving tasks that involve reproducing the whole from a given proper fractional part. More generally, splitting opens new possibilities for developing multiplicative reasoning. The assessment tasks included in this chapter are intended to assess students' construction of the splitting operation and their ability to begin engaging in reversible reasoning with fractions. The instructional tasks build on students' splitting operations to advance their reversible reasoning.

Assessment Task Groups

List of Assessment Task Groups

A7.1: Splitting

A7.2: Reversing Proper Fractions with Rods

A7.3: Reversing Proper Fractions with Strips

A7.4: Reversible Multiplicative Reasoning

(Continued)

(Continued)

Task Group A7.1

Splitting

Materials: fraction strips (five long, thin rectangular strips of paper, all the same size), tape, scissors, writing utensils.

What to do and say: Give the student the fraction strips. Point to one of the strips and tell them that the strip belongs to a student named Tomiko. Tell the student that Tomiko's strip is five times the size of the one belonging to Raphael (another student). Ask the student to make Raphael's strip by using the other copies.

Notes

- The language of this task may be unfamiliar to students, so the teacher might need to emphasize that the *given* strip (Tomiko's) is five times as big as the strip the student should make (Raphael's).
- Students who split should be able to visualize the correct-sized piece by imagining five equal parts in the given strip and cutting it off of a copy. They may want to make marks on the strip before cutting.
- Students do not need the tape to solve the task, but it should be provided for the purpose of assessment. Providing the student with scissors and not tape would suggest to them that the desired strip is smaller than the given one.
- Students who do not split will often produce a strip that is five times as long as the given one. They might do this by taping together five copies of the given strip.
- The teacher should ask the student to justify their response: 'What did you make and how do you know it's what I asked for?'; or 'Can you show me how Tomiko's strip is five times the size of Raphael's strip?'
- This task can be modified to include other numbers, besides 5. However, using the number 2 would not be useful in the assessment because students' work with halves and doubling does not necessarily generalize to other numbers.

Task Group A7.2

Reversing Proper Fractions with Rods

Materials: Fraction rods (also known as 'Cuisenaire rods'; see Appendix for template).

What to do and say: Show the student a dark green rod (six units long) and tell them that it is three-sevenths of the whole. Ask the student to determine how long the whole would be. Similar questions can be asked using other fractions, as long as there is a rod that can represent the unit fraction.

Notes

- In this task and similar tasks, the whole may be bigger than the longest rod (the orange rod), so students might need to put together rods to show how long the whole is.
- In order to solve these tasks, students need to split $\frac{3}{7}$ into three $\frac{1}{7}$ parts. They also need to coordinate three levels of units in activity (the given fraction, the unit fraction and the whole).
- Other good fractions to use for this task include $\frac{5}{8}$ and $\frac{4}{9}$ – non-unit fractions close to $\frac{1}{2}$.
- As in the previous tasks, the teacher should ask the student to justify their response: 'What did you make and how do you know it's what I asked for?'

Task Group A7.3

Reversing Proper Fractions with Strips

Materials: Fraction strips (all the same length), tape, scissors, writing utensils.

What to do and say: Give the student an unmarked fraction strip and tell them that it is three-sevenths of the whole. Ask the student to produce the whole. Similar questions can be asked using any other proper fractions.

Notes

- These tasks should be administered in the same way as those in Task Group A7.2. However, these tasks are more challenging than those in Task Group A7.2 because there are no discrete units of length to check; the student has to produce the lengths, which could be any size.

Task Group A7.4

Reversible Multiplicative Reasoning

Materials: Pencil and paper.

What to do and say: Pose problems of the following kind: '24 ounces is eight times the amount of soda that Stephanie drank; how much soda did Stephanie drink, and can you draw a picture to show this?'

Notes

- These tasks can be posed within various contexts and with different numbers, as long as the second number evenly divides the first number.
- The teacher should assess the accuracy of the student's drawing. Are the lengths approximately proportional to the sizes of the two given numbers? Can the student show how they see the eight times in the picture?

(Continued)

(Continued)

- Sometimes students at Stage 2 can solve the problem numerically but have a hard time representing the relationships in a picture. Students at Stage 2 also sometimes conflate the relationships. For example, they draw 24 ounces marked into three equal parts where each part is 8 ounces.
- The teacher can also ask the student to justify their response. Is the student considering the relative sizes of the two given numbers? Is the student able to identify the unit? Does the student use this unit to produce the requested strip?

Instructional Activities

List of Instructional Activities

IA7.1: Two Hidden Units

IA7.2: My Number, Your Number

IA7.3 My Fraction, Your Fraction

IA7.4: Splitting a Part

IA7.5: Reversing Multiplicative Reasoning

Activity IA7.1

Two Hidden Units

Intended learning: Students will learn to coordinate partitioning and iterating actions while explicitly attending to two units associated with a given non-unit fraction (the unit fraction and the whole). This will also help students learn to reason reversibly with proper fractions – producing the whole from a given proper fraction of it.

Instructional mode: Game that can be demonstrated with the whole class and then played in small groups.

Materials: Fraction rods (see Appendix for template), two envelopes labeled 'whole' and 'unit fraction'.

Description: One student secretly chooses two rods and determines what fraction the smaller rod is of the longer rod. Then they find the associated unit fractional rod. They place the whole rod in an envelope labeled 'whole' and the unit fractional rod in an envelope labeled 'unit fraction'. Then they show the other rod and name it (a non-unit, proper fraction of the whole). Finally, they ask the other students to figure out what rods are in the two envelopes. For example, the student could choose the dark green rod and the orange rod, where the dark green rod is three-fifths of the orange rod. Then they identify the red rod as one-fifth. They place the orange rod in the envelope labeled 'whole' and the red rod in the envelope labeled 'unit fraction'. Then they show the other students the dark green rod, call it three-fifths, and ask the students to figure out the whole and the unit fraction.

Responses, Variations and Extensions

- This activity can be scaled down by having the students choose unit fractions and hiding only the whole.
- The activity can be scaled up by using fraction strips instead of fraction rods. The student who hides the units would have to use a ruler and scissors to make the three strips. This activity is more difficult because fraction strips are continuous in the sense that they can be partitioned into parts of any size. With fraction rods, students can associate these with discrete whole numbers (e.g. the red rod is 2).
- Teachers can also scale up the activity by encouraging students to choose improper fractions.

Activity IA7.2

My Number, Your Number

Intended learning: Students will learn to use splitting in service of a reversible reasoning goal involving whole numbers. They will learn to produce any whole number length from any given whole number length by splitting the given length into units of 1 and then iterating one of those units to produce the desired length.

Instructional mode: Activity for pairs of students.

Materials: Unmarked fraction strips (see Appendix for template), pencil, scissors, tape, ruler.

Description: One student chooses a whole number length and produces a paper strip of that length (with no marks), using a ruler. Then that student asks the second student to produce a strip of a given size, without using the ruler. For example, the first student might make a strip that is 11 centimetres long and ask the second student to make one that is 8 centimetres long. The second student has to make the specified strip using only the given strip, pencil, scissors and tape (no ruler). The pair of students can check the accuracy of the new strip by measuring with the ruler.

Responses, Variations and Extensions

- This activity involves the use of reversible reasoning with whole numbers. Students can rely on their splitting operations to produce a unit of 1 from the given strip.
- Any two whole numbers can be used in this activity, but it's best if neither number is a multiple of the other. Otherwise, students might not need to split or iterate to solve the task.

Activity IA7.3

My Fraction, Your Fraction

Intended learning: Students will learn to use reversible reasoning in service of a goal to produce new fractions. They will learn to produce any proper fraction from any given proper fraction by reproducing the whole and then making a fraction of that whole.

(Continued)

(Continued)

Instructional mode: Activity for pairs of students.

Materials: Fraction strips, pencil, scissors, tape.

Description: One student chooses a fraction and produces that fraction from a fraction strip, being careful to hide any markings they might make in producing the fraction. Then that student asks the second student to produce a different fraction. For example, the first student might make a strip that is three-fourths of a whole strip and ask the second student to make one that is two-thirds of the whole strip. The second student has to make the specified strip and justify their response. The pair of students will have to agree whether the second student produced the correct fraction.

Responses, Variations and Extensions

- This activity requires students to use reversible reasoning with fractions. Students generally have to reproduce the whole from the given fraction in order to make the new fraction.
- Students can choose any proper fractions, but it is best if they avoid unit fractions for the first fraction. Otherwise, the second student might not need to use splitting.

Activity IA7.4

Splitting a Part

Intended learning: Students will learn to produce smaller unit fractions by splitting a given unit fraction. This involves splitting a part and determining the resulting fractional size relative to the whole.

Instructional mode: Activity for pairs of students.

Materials: Unmarked fraction strips, pencil, scissors.

Description: Each pair of students produces a unit fraction from a fraction strip by partitioning it into equal parts and cutting off one of them (e.g. partitioning the strip into five equal parts and cutting off one of them to produce $\frac{1}{5}$). Then one student secretly produces a unit fraction of that unit fraction in the same way (e.g. $\frac{1}{3}$ of the $\frac{1}{5}$ piece). The student then determines the fractions name of the new piece ($\frac{1}{15}$) and tells the second student that name (without showing the new piece). The second student has to produce the new piece ($\frac{1}{15}$) from one of the original parts ($\frac{1}{5}$). The students can then compare the sizes of the pieces they produced.

Responses, Variations and Extensions

- Students can solve these tasks like they did in Activity IA7.3. That is fine, but they should begin to realize that they do not need to make the whole in order to make the specified unit fraction.

Activity IA7.5

Reversing Multiplicative Reasoning

Intended learning: Students will learn to use splitting in order to solve tasks that involve reversible multiplicative reasoning.

Instructional mode: Class activity for students to solve individually.

Materials: Pencil and paper.

Description: Pose problems of the following kind: '$42 is three-sevenths of Sam's money. How much money does Sam have?' Ask students to make drawings to represent their solutions.

Responses, Variations and Extensions

- This task is like assessment task A7.5, except that it includes a proper fraction. Students have to determine the unit fraction ($\frac{1}{7}$) of the composite unit ($42) by splitting the whole into three equal parts. Then students need to coordinate a third level of units, in activity, in order to determine the whole from which that $\frac{1}{7}$ part comes. These tasks involve a more advanced form of reversible multiplicative reasoning (Hackenberg, 2010).

- The context and numbers used in the task may vary, as long as the numerator of the proper fraction is a factor of (evenly divides) the composite unit (e.g. 3 divides 42). It is best to use whole numbers that are divisible by both the numerator and denominator of the fraction (i.e. 42 is divisible by both 3 and 7) so that students have to determine the number of equal parts in which to divide the whole number.

- As an extension, teachers can choose numbers that do not work out to whole number answers: For example, '$37 is two-fifths of Sam's money; how much money does Sam have?'

- The context for these problems does not have to be money. For example, '42 feet is three-sevenths of the distance from the door to the playground. How far is the distance from the door to the playground?' Using distance contexts can encourage linear or rectangular drawings of the relationship. Linear or rectangular drawings can also be used to model the money situation, and they are often helpful here.

- Students who have not learned to reverse their multiplicative reasoning in this way may try to find 3/7 of $42. For those students, the task can be scaled back by using a unit fraction (e.g. $\frac{1}{7}$) and a smaller composite unit (e.g. $10 or $21).

- These activities can be used to support standards like Common Core State Standard 5.NF.4, which involves multiplying a whole number by a fraction.

- These activities also support complexifying, by having students consider fractions of composite units, rather than simple wholes.

8

Teaching Students at Stage 3: Fractions as Numbers

Students at Stage 3 can take the coordination of three levels of units as given. Recall in Chapter 3 we discussed that this means they can 'see' numbers as structured into three levels of units prior to any work in solving a problem. This does not mean they automatically know that, for example, 35 is 5 units of 7. However, they anticipate that 35 (or any number) can be structured in such a way.

Researchers have found that this ability to coordinate three levels of units means students can develop fractions as multiples of unit fractions (Hackenberg, 2007; Steffe and Olive, 2010). For example, for these students $\frac{3}{7}$ is three times $\frac{1}{7}$, and $\frac{7}{5}$ is seven times $\frac{1}{5}$. This meaning for fractions is critical because, as we have said in Chapter 6, a part–whole meaning for $\frac{7}{5}$ does not make sense. Furthermore, when fractions are multiples of unit fractions they become 'numbers in their own right' for students (Hackenberg, 2007). When any fraction is a multiple of a unit fraction, students can develop virtually all arithmetical operations with fractions based on conceptual meaning (Steffe and Olive, 2010).

This chapter focuses on how students at Stage 3 can learn to conceive of fractions, how teachers can assess and promote this way of thinking about fractions, and how they can work with students at Stage 3 to further solidify their concepts of fractions. In Chapter 9 we address the fifth level of fragmenting that is associated with students at Stage 3, and in Chapters 10, 11 and 12 we address arithmetical operations with fractions for students at Stages 2 and 3.

The Iterative Fraction Scheme: Fractions as Numbers

The name researchers give to conceiving of any fraction as a multiple of a unit fraction is the *Iterative Fraction Scheme*. For a student with this scheme, any fraction, even an improper fraction, is meaningfully made by iterating a unit fraction. The student thinks of a fraction like $\frac{7}{5}$ as $\frac{1}{5} \times 7$ *and* as 1 unit ($\frac{5}{5}$) and $\frac{2}{5}$ of that unit.

To think this way, the student has to see $\frac{7}{5}$ as a unit of 7 units (fifths). However, any of these one-fifths could be iterated five times to produce 1, the whole from which everything was made (see Figure 8.1). So, there are three levels of units that the student coordinates: the unit fraction ($\frac{1}{5}$ in the example below), the whole from which the unit fraction was made ($\frac{5}{5}$ below) and the fraction itself ($\frac{7}{5}$ below). Keeping track of these three related units allows students to create the two meanings for $\frac{7}{5}$: $\frac{1}{5} \times 7$, as well as 1 and $\frac{2}{5}$.

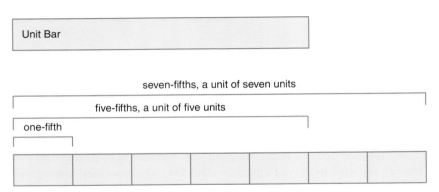

Figure 8.1   The three levels of units involved in thinking of $\frac{7}{5}$ as $\frac{1}{5} \times 7$ *and* 1 and $\frac{2}{5}$

Assessing an Iterative Fraction Scheme

An important way to assess whether students have developed an Iterative Fraction Scheme is to pose the following task:

'Drawing Seven-Fifths Problem': The rectangle below shows a candy bar. There is another candy bar that is seven-fifths of that bar. Can you draw the other bar?

Students at Stage 3 may find this to be a tough or unusual problem – most students will not have encountered it before. Because of their ability to coordinate three levels of units, they will often draw a picture like the one in Figure 8.2 and explain that they thought of drawing a new bar consisting of a copy of the original bar and two more fifths of the original bar.

Figure 8.2   Typical response from a student at Stage 3 to the 'Drawing Seven-Fifths Problem'

Then it is helpful to ask questions about the pictures to find out more about students' meanings. Here are some useful follow-up questions:

a.   What is the fractions name of the bar you drew?
b.   [If students did not draw all the parts in the new bar] Can you show all the parts in your new bar?
c.   How did you know how long to make the new bar?
d.   [Teacher shades the sixth or seventh part in the new bar.] What fraction name do you give to that part?
e.   Shade one-fifth. [Students will often ask in which bar, and the teacher can say 'anywhere'.]

Question (a) might seem redundant, since the student was just asked to draw $\frac{7}{5}$. However, after drawing the bar students who lack an Iterative Fraction Scheme will sometimes call the bar $\frac{7}{7}$ because they lose track of the whole when they iterate the unit fraction past the whole (Tzur, 1999). Thus, they make the new, longer bar the whole.

Students who have developed an Iterative Fraction Scheme will usually be careful in drawing and aligning parts. For example, in Figure 8.2 the first five parts in the new bar align pretty closely with the five parts in the original bar. This is one indication that the students think of the parts in the original bar as identical to the parts in their new bar – i.e. they are all the same size and interchangeable. For a response to (c), students who have developed an Iterative Fraction Scheme often describe that they knew $\frac{7}{5}$ was a whole bar and two more fifths, and so they drew a whole bar and then two more fifths of that bar.

To investigate further how they see those seven parts, (d) is helpful, because for some students – even students at Stage 3 – once they draw a bar with seven equal parts in it, the parts become sevenths. Students who say that the part is one-fifth of the original bar and one-seventh of the new bar show the double-naming in relation to different referent units that also characterizes students at Stage 3. Part (e) can be asked instead of part (d), as described in the next section. Alternately, if students do respond 'one-seventh' to part (d), then the teacher might pose part (e) to find out more about their thinking about where the fifths are in the drawing.

Promoting an Iterative Fraction Scheme: Magic Candy Bars

The first author has worked with two 6th-grade students, Bridget and Deborah, who found the 'Drawing Seven-Fifths Problem' challenging (Hackenberg, 2007). They each drew a picture similar to Figure 8.2 but without all the parts marked. They each explained that they had drawn a bar the same size as the original but added two parts. When asked to show the parts, they marked five parts in the top bar and seven in the bottom bar, but all the parts were not aligned (Figure 8.3).

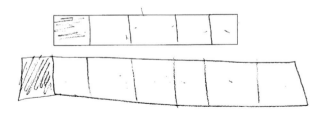

Figure 8.3   Bridget's initial picture for the 'Drawing Seven-Fifths Problem'

Source: Hackenberg, 2007

A teacher asked them to 'colour one-fifth'. He purposefully did not specify where. The students asked where, and he said, 'anywhere'. They each coloured the first part of the 5-part bar. Then he asked, 'What about those pieces in the bottom candy bar?' Both students said that the pieces became sevenths since there were seven parts. When asked to shade a part in the lower bar and tell how much it was of the top bar, both students said, 'uh oh!', thought for awhile and then reported being stumped.

The teacher had assessed that both had developed at least a Measurement Scheme for Unit Fractions (see Chapter 6) and that they could make new fractions by iterating a unit fraction, at least within the whole. So, he asked each student to start with a fresh copy of the bar. He told them that this bar was a 'magic' candy bar, in that when they took out a piece, it would fill right back in. Having the idea of a magic bar can allow students to feel free to take as many pieces of it as they want – rather than only the number of parts they get (e.g. five) when they partition.

The teacher asked the students to draw a bar that was three-fifths of the given bar. Each student marked the original bar into five equal parts and drew another bar spanning three of those parts. When asked how they knew that they had drawn three-fifths, each student marked in the three parts (Figure 8.4, middle bar). The teacher asked them to shade a part in that bar and tell how much it was of the top bar, and without hesitation both students stated that it was one-fifth of the top bar.

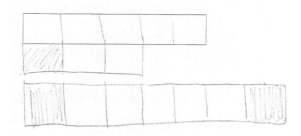

Figure 8.4   Bridget's pictures of three-fifths and seven-fifths of the given bar

Source: Hackenberg, 2007

Then the teacher asked them to make a bar consisting of seven of those parts. Bridget's drawing is shown (Figure 8.4, bottom). The teacher asked each student to shade the first part and tell how much it was of the top bar. The students did so and said promptly and softly, 'one-fifth'. Bridget explained that the parts in the bottom bar were equal to the parts in the middle bar and top bar. Both students also stated that the last part in the bottom bar was one-fifth of the top bar, and they named the new bar 'seven-fifths'.

Now, as it turned out, one of these students was at Stage 3 and one was an advanced Stage 2 student. The Stage 3 student went on to demonstrate the use of an Iterative Fraction Scheme throughout the semester, while the Stage 2 student did not. These observations contributed to the finding about the importance of being a Stage 3 student in developing fractions as numbers (Hackenberg, 2007; Steffe and Olive, 2010).

Promoting an Iterative Fraction Scheme: Fraction Comparisons and Drawing Longer Bars

Another way to promote the development of an Iterative Fraction Scheme is to pose a variety of fraction comparison problems with fractions that are close to 1 (a little smaller or a little larger). For example, students can play a 'Fraction Comparison Partner Game'. The teacher gives or shows a fraction to each partner privately. Each student draws their fraction based on a common bar that is their 'unit', and each hides their bar under a cover (e.g. a piece of paper). Then the students tell each other their fractions and determine (without looking at the drawings) which is bigger. Once they predict, they remove the covers and compare the drawings to check (1) whether the drawings are correct, and (2) whether their predictions are correct. Fractions that can be used include fractions that are the same number of unit fractions smaller than the unit (e.g. $\frac{7}{8}$ v. $\frac{8}{9}$, $\frac{9}{11}$ v. $\frac{11}{13}$) and fractions one unit fraction larger than the unit (e.g. $\frac{4}{3}$ v. $\frac{5}{4}$, $\frac{11}{10}$ v. $\frac{8}{7}$). The activities at the end of this chapter elaborate on this game.

In addition, it can be helpful to engage students in a variety of problems about drawing bars that are longer than a given bar. Consider the following problems:

'Specific Longer Candy Bar Problem': Draw a candy bar that will be your unit. Draw a copy of the unit, and make that copy into $\frac{8}{8}$, which we will call an $\frac{8}{8}$-bar. Draw a bar $\frac{2}{9}$ longer than the $\frac{8}{8}$-bar. How long is the bar you drew? How do you know?

'Open Longer Candy Bar Problem': Draw a bar that will be your unit. Draw a copy of the unit, and make that copy into $\frac{11}{11}$, which we will call an $\frac{11}{11}$-bar. Use thirteenths to draw a bar a little bit longer than the $\frac{11}{11}$-bar. How long is the bar you drew? How do you know?

To solve each of the above problems, students have to think about how many of the target fractions make a whole unit ($\frac{9}{9}$ or $\frac{13}{13}$) and then extend beyond that bar. In the 'Specific Longer Candy Bar Problem' students may call the bar 'eleven-elevenths' if they are not taking the $\frac{9}{9}$ as a unit, in addition to the 11 parts as a unit in relation to the $\frac{9}{9}$. In the 'Open Longer Candy Bar

Problem' students may wonder if the problem can have multiple answers (it can), which can prompt interesting discussions about what it means to be 'a little bit longer'.

We note that, in addition to what we have discussed in this section, engaging in numerous tasks that involve partitioning and iterating is beneficial for students in the development of an Iterative Fraction Scheme. Tasks from prior chapters as well as tasks in Chapters 9, 10, 11 and 12 can be helpful.

Tasks to Solidify Conceiving of Fractions as Numbers in Their Own Right

Once students have developed an Iterative Fraction Scheme, there are several ways to reinforce their work with fractions as multiples of unit fractions and with the involved coordinations of units. We give two examples below, but there are more in Chapters 9 and 10.

> Making a Fraction Number Sequence: Start with the candy bar given; it is the unit. Make $\frac{1}{5}$ of the unit, which we will call a $\frac{1}{5}$-bar. Now make a $\frac{2}{5}$-bar. Continue to make a sequence of fraction bars for fifths, and go up to at least $\frac{11}{5}$. This sequence is called a fraction number sequence. [The follow-up questions below could be posed verbally in discussion.]
>
> **a)** How many numbers are in the fraction number sequence of fifths?
>
> **b)** How is this fraction number sequence like the sequence of whole numbers? How is it different from the sequence of whole numbers?
>
> **c)** How many fraction number sequences are there?

Making fraction number sequences for many different 'fraction families' (e.g. thirds, fourths, ninths, twelfths) is important because it can allow students to see that fractions continue indefinitely just like whole numbers. However, fraction number sequences are different from the sequence of whole numbers because every so many there is a special fraction that is also a whole number.

Another way to help students solidify conceiving of fractions as numbers is to play the 'Fraction Comparison Partner Game' with fractions that are quite a bit larger than the unit. For example, partners could compare $\frac{21}{5}$ and $\frac{17}{4}$. Making this comparison requires determining the number of $\frac{5}{5}$s contained in $\frac{21}{5}$, and the number of $\frac{4}{4}$s contained in $\frac{17}{4}$. Since both numbers contain four whole units, and there is one more unit fraction contained in each number, $\frac{17}{4}$ must be bigger because $\frac{1}{4}$ is bigger than $\frac{1}{5}$. As students make progress with these comparisons, increasing the number of units in the number, decreasing the size of the parts used, and increasing the number of unit fractions the number is beyond a whole number of units are all ways to augment the level of difficulty. For an example, consider comparing $\frac{125}{11}$ with $\frac{103}{9}$; it may be quite a challenge! The activities at the end of this chapter elaborate on this version of the game.

Assessment Task Groups

List of Assessment Task Groups

A8.1:   Drawing 'Large' Fractions

A8.2:   Drawing the Whole Given a Fraction Larger than the Whole

A8.3:   Drawing a Mixed Number

A8.4:   Magic Candy Bars

A8.5:   Drawing a Longer Fruit Strip

A8.6:   Which Is Bigger?

Task Group A8.1

Drawing 'Large' Fractions

Materials: Multiple copies of a piece of paper with a rectangle on it.

What to do and say: Tell the students that the rectangle is an energy bar. Tell them also that there is another energy bar that is seven-fifths the size of that bar [do not write down fraction notation, just say this verbally]. Can they draw the other bar? When they draw another bar, ask the students to explain how they know their bar is seven-fifths of the first bar. The four follow-up questions for the 'Drawing Seven-Fifths Problem' in the chapter are useful here as well:

a)   [If students did not draw all the parts in the new bar] Can you show all the parts in your new bar?

b)   How did you know how long to make the new bar?

c)   [Teacher shades the sixth or seventh part in the new bar.] What fraction name do you give to that part?

d)   Shade one-fifth. [Students will often ask in which bar, and the teacher can say 'anywhere'.]

Notes

*   Students who have not developed an Iterative Fraction Scheme will have a variety of responses to this problem. Some will say it cannot be done because it does not make sense to draw 7 parts out of 5. If they do, you might try Task Group A8.4 to see if having a 'magic' candy bar will alleviate that issue.

*   Some students will draw $\frac{5}{7}$ in an attempt to coordinate 7 parts and 5 parts. If they do, you can press them about whether their bar shows $\frac{7}{5}$. We have also witnessed students draw 75% of the bar in an attempt to include the idea of 7 and 5 in the problem.

(Continued)

(Continued)

- Some students may not have reasons for why they drew the bar a little bit longer or a lot longer (or a little bit shorter or a lot shorter). If they don't, it is best not to press too hard but you might move to Instructional Activity 8.1.
- If students are successful at this task, try Task Group A8.2.

Task Group A8.2

Drawing the Whole Given a Fraction Larger than the Whole

Materials: Multiple copies of a piece of paper with a rectangle on it (different in length from the one for A8.1).

What to do and say: Tell the students that the rectangle is an energy bar. Tell them that this bar is nine-sevenths the size of another energy bar [do not write down fraction notation, just say this verbally]. Can they draw the other bar? When they draw another bar, ask these follow-up questions:

a) [If students did not draw all the parts in the new bar] Can you show all the parts in your new bar?
b) How did you know how long to make the new bar?

Notes

- Students can succeed at Task A8.1 and not succeed at this task. That is, this task requires students to reverse their reasoning in ways similar to those described in Chapter 7. To solve the task a student has to see the given bar as 9 times $\frac{1}{7}$ of the other bar, and so partitioning the bar into 9 equal parts allows the student to identify $\frac{1}{7}$ of the other bar. Once the student has identified $\frac{1}{7}$ of the other bar, they can iterate that amount 7 times to make the whole bar. If students do not succeed at this task, it casts doubt on how solidly they have coordinated the different levels of units involved in making fractions as numbers.
- If students do succeed at this task, it is a good confirmation that they are coordinating the different levels of units involved in making fractions as numbers as described in Chapter 8.

Task Group A8.3

Drawing a Mixed Number

Materials: Blank paper.

What to do and say: Ask the students to draw a picture of a mixed number such as $4\frac{7}{9}$. After they do so, ask them to explain their drawing.

Notes

- Students may draw the 4 and $\frac{7}{9}$ as not connected to each other. For example, they may draw four circles and then a bar that they partition into nine parts, shading seven of them. If so, then that is good evidence that students do not see the ninths as connected to the whole units that make up the 4. This kind of a response indicates that students are not coordinating the different levels of units involved in making fractions as numbers as described in Chapter 8. A possible follow-up is: 'I noticed that you drew circles for the 4 here. What does the rectangle represent?'

Task Group A8.4

Magic Candy Bars

Materials: Multiple copies of a piece of paper with a rectangle on it.

What to do and say: The candy bar shown below is 'magic'. That means when you take a piece out, it fills right back in. Use the magic candy bar to make a candy bar that is seven-fifths of that bar – you can draw your new bar.

Notes

- As discussed, this problem may help to alleviate students' concern that there is not enough candy to make seven parts – that is, there are only five parts.

Task Group A8.5

Drawing a Longer Fruit Strip

Materials: Multiple copies of a piece of paper with a rectangle on it, and a copy of that rectangle marked into fifths.

What to do and say: Here is a fruit strip, and a copy of the strip. The copy is marked into five equal parts, so we'll call it a $\frac{5}{5}$-strip. Draw a strip that is $\frac{1}{7}$ longer than the $\frac{5}{5}$-strip. How long is the strip you drew? How do you know?

Notes:

- Task Group A8.5 can be used if students were not successful with A8.1. However, it is a more complex problem statement than A8.1, and so the students who struggled with A8.1 may find it tough.

(Continued)

(Continued)

Task Group A8.6

Which Is Bigger?

Materials: Multiple copies of a piece of paper with a rectangle on it.

What to do and say: Give the student two fractions just larger than one, such as $\frac{4}{3}$ and $\frac{5}{4}$. Ask the student to predict which is bigger and why. Then the student should draw both in relation to the given rectangle and compare.

Notes

- A8.6 should be used only if students have been successful with A8.1.
- Using *JavaBars* software here is helpful (see math.coe.uga.edu/olive/welcome.html).

Instructional Activities

List of Instructional Activities

IA8.1:   Magic Candy Bars

IA8.2:   Fraction Comparison Game

IA8.3:   Drawing a Longer Fruit Strip

IA8.4:   Making a Fraction Number Sequence

Activity IA8.1

Magic Candy Bars

Intended learning: This activity is intended to support students' development from a Measurement Scheme for Unit Fractions toward an Iterative Fraction Scheme.

Instructional mode: Game to be played in pairs.

Materials: Unmarked fraction strips (see Appendix) all the same size, pencil, scissors, tape.

Description: Students will be given a 'magic candy bar' and several copies of it (unmarked fraction strips). One student plays the role of customer, and the other plays the role of storekeeper. The customer can ask for any size candy bar they would like, and the storekeeper has to make that candy bar using copies of the magic candy bar. For example, if the customer orders $\frac{6}{5}$ of the magic candy bar, the storekeeper can partition a copy of the magic candy into five equal parts, cut off one of those parts, make a total of six copies of that part, and tape them together. The customer has to decide whether the order was fulfilled correctly. The students should switch roles after each completed order.

Responses, Variations and Extensions

- Customers should order proper fractions at the start but can extend the range of numbers by ordering improper fractions when the storekeeper seems ready for the additional challenge.
- In line with standards like Common Core State Standard 4.NF.4, this activity can help students understand fractions – even improper fractions – as a multiple of a unit fraction.

Activity IA8.2

Fraction Comparison Game

Intended learning: Students will learn to compare the sizes of fractions.

Instructional mode: Game to be played in pairs.

Materials: A fraction strip, dice, paper and pencils.

Description: Students are given a fraction strip as a referent whole throughout the game. Each student rolls a pair of dice and uses the two numbers to make a fraction, where the smaller number is the numerator. The students draw their respective fractions relative to the size of the whole unit (the given fraction strip) and cover them with another sheet of paper. Then they tell each other their fractions and determine (without looking at the drawings) which fraction is bigger. Once they predict, they remove the covers and compare the drawings to check (1) whether the drawings are correct, and (2) whether their predictions are correct.

Responses, Variations and Extensions

- The covers are used to distance the instructional setting, so that students begin to imagine the sizes of fractions without looking at them.
- Students can extend the range of numbers in this game by rolling the dice twice to determine their fractions. For example, if a student rolled a 7 and 11 (by rolling the pair of dice twice), their fraction would be $\frac{7}{11}$.
- Later (preferably after the students have mastered activities IA8.3 and IA8.4) they can use the larger number for the numerator so that they compare improper fractions.
- This activity can help students achieve standards for comparing fractions of different numerators and denominators, like Common Core State Standard 4.NF.2.
- Note that activity IA6.1 addressed comparing unit fractions. Activity IA8.2 involves complexifying these comparisons by including non-unit fractions (even improper fractions).

(Continued)

(Continued)

Activity IA8.3

Drawing a Longer Fruit Strip

Intended learning: Students will learn to coordinate the whole unit and unit fractions while making an improper fraction.

Instructional mode: Whole-class activity.

Materials: Marked and unmarked fraction strips (see Appendix), pencil, scissors, tape.

Description: The teacher will give each student an $\frac{8}{8}$-bar (fraction strip marked into eight parts). The teacher will ask the students to make a bar that is $\frac{2}{9}$ longer than the $\frac{8}{8}$-bar. Then the teacher can ask follow-up questions: How long is the bar you drew? How do you know?

Responses, Variations and Extensions

- To solve these types of problems, students have to think about how many of the target fractions make a whole unit. For example, $\frac{9}{9}$ makes the whole, so the fraction that is 2/9 longer than $\frac{8}{8}$ will be $\frac{11}{9}$.
- Variations of the problem include asking students to make a bar that is 'just a little longer' than the $\frac{8}{8}$-bar and to name it. Some students might make $\frac{9}{8}$, but collectively, they should realize there are many possible answers, such as $\frac{10}{9}$ or $\frac{12}{11}$.
- The teacher can also vary the task by giving students other marked fraction strips besides 8/8.

Activity IA8.4

Making a Fraction Number Sequence

Intended learning: Students will generate proper and improper fractions by iterating a unit fraction, thereby supporting their understanding of fractions as numbers on a number line.

Instructional mode: Whole-class activity.

Materials: Unmarked fraction strips, paper, pencils.

Description: The teacher will give each student a fraction strip, which will serve as the referent whole unit. Then they will ask students to draw $\frac{1}{5}$ of the unit, calling this a $\frac{1}{5}$-bar. They will then ask students to use this unit to produce a $\frac{2}{5}$-bar, $\frac{3}{5}$-bar and so on, up to $\frac{12}{5}$. The sequence of bars is called a fraction number sequence. The teacher can begin to ask the students questions about this sequence:

1) How many numbers are in the fraction number sequence of fifths?
2) How is this fraction number sequence like the sequence of whole numbers? How is it different from the sequence of whole numbers?
3) How many fraction number sequences are there?

Responses, Variations and Extensions

- Students should begin to realize that there are an infinite number of fraction number sequences and that each one contains an infinite number of fractions. This realization involves structuring numbers.

- The teacher should continue the activity with other fraction number sequences, such as thirds or sevenths.

- This activity supports students' understanding of fractions as numbers on a number line – a concept that relates to standards that span multiple grade levels (e.g. Common Core State Standards 3.NF.2 and 6.NS.6).

9

Teaching Students at Stages 2 and 3: Equal Sharing of Multiple Items

Domain Overview

After sharing single items equally, a next step is to share multiple identical and non-identical items equally. For example, sharing two identical granola bars equally among three people is a difficult problem for many elementary school students, but it opens the way to think about the size of one share in relation to a single bar, and in relation to both bars. Equal sharing problems at varying levels of difficulty (e.g. Empson and Levi, 2011) are useful in helping students develop their reasoning about sizes of shares, and the parts that make up those shares, as well as fraction language and notation. In our view, students need to be at least at the third level of fragmenting in order to make progress on these problems. In this chapter we provide:

(1) A discussion of an advanced solution to an advanced equal sharing of multiple items problem in order to demonstrate the fifth level of fragmenting that corresponds to Stage 3 students.

(2) A series of equal sharing of multiple items problems from less difficult to more difficult with a discussion of student solutions to these problems in order to demonstrate the difficulty range of the problems.

Equally Sharing Multiple Items and the Fifth Level of Fragmenting (Stage 3 Students)

Karenna works on the following problem:

'Five Cereal Bars Problem': Share 5 identical cereal bars (rectangles) equally among 7 friends. Show how you made the equal shares. Draw out the share for one friend separately.

1) How much is one friend's share when compared with one cereal bar? Explain.

2) How much of all the bars does one friend get? Justify your answer.

3) How can you convince someone that the amount you have drawn is an equal share? Explain.

To solve the problem, Karenna partitions each rectangle into seven equal parts. Then she shades the first part of each of the five bars (Figure 9.1, *top*). She draws out these five parts, connected, as the share for one person (Figure 9.1, *bottom*). In explanation she says, 'To make the shares I marked each bar into seven equal parts. Then one person gets one part from each part. If you put those together, that's the share for one person.'

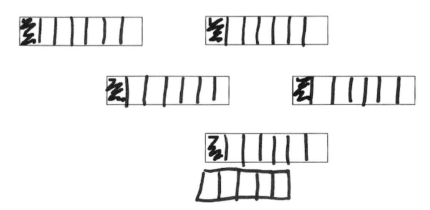

Figure 9.1   Karenna's solution to the 'Five Cereal Bars Problem'

She goes on to say, 'One person's share is five-sevenths compared to a cereal bar, because it consists of five one-sevenths – the size of the parts are sevenths because I marked each bar into seven equal parts.'

'But, one person's share is one-seventh of all the bars because seven friends are sharing all the bars. If they do it equally, one share has to be one-seventh. Also, five parts out of 35 parts total is the same as one-seventh of the total.'

In justifying that she had made an equal share, Karenna explains, 'I know the share is an equal share because if you take the five-part share and repeat it seven times, you get 35 of those parts, which is all of the bars. That means that the share is exactly one-seventh of all the bars.'

Karenna's solution is representative of a student at the fifth and most advanced level of fragmenting (Steffe and Olive, 2010). We present this solution to indicate what that level is, and what is important to understand about this way of thinking and acting. In particular, we highlight three important features of Karenna's thinking: distributive reasoning, working with multiple levels of units and iterating.

First, an important feature of Karenna's thinking is that she engaged in *distributive reasoning*. She had a goal to share 5 bars equally among 7 friends, and to accomplish that goal she shared each of the 5 bars among the 7 friends. If we were to notate this solution with division, we might write:

$$5 \div 7 = (1 + 1 + 1 + 1 + 1) \div 7 = (1 \div 7) + (1 \div 7) + (1 \div 7) + (1 \div 7) + (1 \div 7) = \frac{1}{7} + \frac{1}{7} + \frac{1}{7} +$$
$$\frac{1}{7} + \frac{1}{7} = \frac{5}{7}$$

This notation highlights the use of a version of the Distributive Property. (Note that in the Distributive Property of Division, the divisor can be distributed over the dividend, but not vice versa.) We do not claim that Karenna herself would write this notation, just that as teachers we can write it to further understand how she distributed the process of sharing (dividing) across each bar in order to share the entire collection of bars. Alternately, the following notation would highlight taking one-seventh of 5 bars by taking one-seventh of each of the bars:

$$\frac{1}{7} \times 5 = \frac{1}{7} \times (1 + 1 + 1 + 1 + 1) = (\frac{1}{7} \times 1) + (\frac{1}{7} \times 1) + (\frac{1}{7} \times 1) + (\frac{1}{7} \times 1) + (\frac{1}{7} \times 1) = \frac{1}{7} + \frac{1}{7}$$
$$+ \frac{1}{7} + \frac{1}{7} + \frac{1}{7} = \frac{5}{7}$$

This notation demonstrates the idea that $\frac{1}{7}$ of 5 is the same as $\frac{5}{7}$ of 1, and that the size of the share in comparison with 1 bar is $\frac{5}{7}$. This important idea about fractions of whole numbers can be generalized. For example, if I want to find $\frac{1}{32}$ of 17, the result must be $\frac{17}{32}$ because it is $\frac{1}{32}$ from each 1 that makes up 17, or $\frac{1}{32}$ 17 times. That is, we can abbreviate or compress the distributive reasoning involved in taking $\frac{1}{32}$ of 17 as follows:

$$\frac{1}{32} \times 17 = \frac{1}{32} \times (1 \times 17) = (\frac{1}{32} \times 1) \times 17 = \frac{17}{32}$$

This thinking represents the use of the Associative Property of Multiplication, and we call it *associative reasoning*. (Of course, this notation also demonstrates the Multiplication Identity Property.) Again, we do not suggest that a young student would necessarily write this notation, but that it helps teachers to see how it is possible to compress the notation that reflects distributive reasoning in efficient ways.

Distributive reasoning and associative reasoning are important for developing schemes for multiplying fractions (Hackenberg and Tillema, 2009; Steffe and Olive, 2010; Tillema and

Hackenberg, 2011), as we address in Chapter 10. In addition, both ways of reasoning are valuable in many mathematical domains, including whole numbers.

A second important feature of Karenna's thinking is that she had two different ways to view the share, a five-part bar, in comparison with all of the bars. She knew that if she was sharing all of the bars equally among seven people, each share must be $\frac{1}{7}$. This means she had developed the idea that no matter the amount, marking that amount into a certain number of equal shares means each share is a unit fraction of the amount. In addition, she had ideas about how $\frac{5}{35}$ was the same as $\frac{1}{7}$; we address the issue of different names for the same-size fraction (commensurate fractions) in Chapter 11.

A third important feature of Karenna's thinking is that she used iteration of a share to justify why the share she had made must be $\frac{1}{7}$ of all of the candy. That is, she had developed the idea that if a share is a unit fraction of an amount, that share can be repeated a whole number of times to make the total amount. Since the five-part bar repeated seven times produced 35 parts, all of the cereal bars, it must be $\frac{1}{7}$ of all of the bars.

Equally Sharing Multiple Items: Using Repeated Halving

Many elementary school students do not solve the 'Five Cereal Bars Problem', or other equal sharing of multiple items problems, as Karenna did (Empson and Levi, 2011; Hackenberg and Lee, in press; Lamon, 1996; Steffe and Olive, 2010). So, here we address other key ways of thinking that students will demonstrate in solving equal sharing of multiple items problems.

Problems easier than the 'Five Cereal Bars Problem' include those for which students might not encounter the need to use distributive or associative reasoning. For examples, consider equally sharing 2 bars among 4 people, 3 bars among 4 people, and 7 bars among 8 people. Students do not have to use distributive reasoning to solve any of these equal sharing problems because they can use 'repeated halving', that is, halves, fourths and eighths, to come to a solution.

For example, let's say that a student is going to share 3 cereal bars equally among 4 people. The student could partition each bar into halves and give one-half of a bar to each person. Then the student could partition the last two halves each in half again, thereby making fourths. Each person also gets one-fourth of a bar (Figure 9.2).

Figure 9.2   Sharing 3 bars equally among 4 people using halves and fourths

So, each person's share is one-half of one bar and one-fourth of one bar. Students who make this solution may not yet know a single fraction name for the share, although they might be able to develop one based on the reasoning they did in the problem (two one-fourths of one bar are equal to one-half of one bar).

Equally Sharing Multiple Items: Sharing Whole Bars First, and Then Partitioning a Single Bar or Using Repeated Halving

Problems easier than the 'Five Cereal Bars Problem' also include those where students can share whole bars first, and then deal with a single bar that they can share into however many parts necessary. For examples, consider equally sharing 4 bars among 3 people, 6 bars among 5 people, and 9 bars among 4 people. Students do not have to use distributive reasoning to solve these kinds of equal sharing problems because they only need to partition a single bar if they first give out whole bars.

For example, let's say that a student is going to share 4 bars equally among 3 people. The student could give out a whole bar to each person, and then there is just one bar left (Figure 9.3).

Figure 9.3   Sharing 4 bars equally among 3 people

The student could then partition that single bar into three equal parts. So, one person's share is one bar and one-third of a bar. We strongly caution that this amount may not also be four-thirds for the student. For example, if asked to draw 1 [mixed number] $\frac{1}{3}$ outside of this problem context, students might draw 1 and $\frac{1}{3}$ as a whole bar and then a bar that is partitioned into three parts where one part is shaded; thus the whole and the thirds may not be related to each other (see Chapter 8).

In this same category we include problems where students can share whole bars first and then engage in using halves and fourths to work with the remaining bars. For examples, consider equally sharing 6 bars among 4 people, 7 bars among 4 people, and 15 bars among 8 people.

Equally Sharing Multiple Items: Bricolage Solutions for Harder Problems

Problems similar to the 'Five Cereal Bars Problem' in difficulty include equally sharing 2 bars among 3 people, 3 bars among 5 people, and 7 bars among 10 people. In all of these problems, the number of bars and number of people are relatively prime, which means there are no common factors (other than 1) between the two numbers. In addition, in all of these problems repeated use of halves and fourths and eighths will not solve the problem: Somewhere along the way a student needs to use some other number of partitions (e.g. thirds, fifths, sevenths) with more than one bar. These problems can promote the use of distributive reasoning.

However, students can also make what we refer to as *bricolage* solutions for these problems. In French, the word *bricolage* refers to tinkering with a diverse set of items that are available or at hand in order to create a product. For example, in art, a bricolage sculpture would be one consisting of found objects put together in a trial-and-error, playful manner. In education, the term has been used to refer to solution processes that involve trying, testing and playing around, in contrast with a more direct analytical process (Turkle and Papert, 1992). We use the term bricolage solutions for equal sharing of multiple items problems to indicate that the student is using available 'tools' of halving and partitioning bars into other parts in order to get to a solution, but the number of parts to use is determined in the moment as the student sees the effect of each round of partitioning and the giving out of parts.

For example, consider the 'Five Cereal Bars Problem'. A student might start by halving each bar, making a total of 10 one-half bars (Figure 9.4).

Figure 9.4   Partitioning each bar into halves and using 7 of the 10 halves

The student gives out a one-half bar to each person and has three one-half bars left. Then the student partitions each of those halves into thirds, making a total of 9 small parts (thirds of halves, or sixths, but the student does not necessarily know that). The student gives out one of these small parts to each person and has two small parts left (Figure 9.5).

Figure 9.5 Partitioning 3 leftover halves each into thirds (small parts), and using 7 of the 9 small parts

The student then partitions each small part into four parts, making a total of 8 mini parts (fourths of sixths, or twenty-fourths, but again the student is unlikely to know that). The student gives out one mini-part to each person and has one mini-part left (Figure 9.6).

Figure 9.6 Partitioning 2 leftover small parts each into fourths (mini-parts), and using 7 of the 8 mini-parts

The student then partitions that last mini-part into 7 equal teeny-parts and gives 1 teeny-part to each person (one-seventh of one twenty-fourth, or $\frac{1}{168}$!). So, one person's share consists of one-half of a bar, a small part, a mini-part and a teeny-part (Figure 9.7).

Figure 9.7 One-seventh of 5 is one-half plus a small part plus a mini-part plus a teeny-part

It is very unlikely that a student will determine a single fraction name for this share. However, they have indeed made the shares!

Equally Sharing Multiple Items: Solutions Involving Multiplication and Division

Some students solve problems like the 'Five Cereal Bars Problem' by using whole number multiplication and division. Although these solutions are not bricolage solutions, they also are not solutions that involve distributive reasoning in the way we have shown Karenna using it.

For example, consider again the 'Five Cereal Bars Problem'. Some students solve this problem by determining that they need a total number of parts in the 5 bars that they can divide by 7. So they start to think about multiples of 5 that will work. Students who are just starting this kind of solution might think about partitioning each bar into 2 equal parts to produce 10 parts total. But 10 is not divisible by 7. Then they might think about partitioning each bar into 3 equal parts to produce 15 parts total. But 15 is not divisible by 7. Then they might realize that if they could make 35 parts total, that would be divisible by 7 – and that would require making 7 parts in each of the bars. So, these students partition each bar into 7 parts to produce a total of 35 parts, and then divide by 7 to get 5 parts in one share. A student's drawing for this solution might not look that different from Karenna's (Figure 9.1), but it would be different in that the student is thinking differently.

This kind of solution can be generalized: students can determine that multiplying the two numbers will give a usable total number of parts, or they might even determine that the least common multiple between the two numbers will produce the smallest usable total number of parts. However, in these solutions students are not necessarily aware of working with fractional parts. That is, it may be a new problem for them to consider the fraction size of one share in relation to one bar or all of the bars because they are not thinking explicitly about the fractional sizes of parts in their solution process. In contrast, in Karenna's solution to the 'Five Cereal Bars Problem', she is clearly aware of working with fractional parts in explicit and efficient ways.

The following task groups are intended to assess students' understanding of each of the solution types illustrated in this chapter. These assessments can be used to determine their sophistication in sharing multiple items equally. We then share instructional activities designed to promote greater sophistication. The assessment tasks and instructional activities occur in four different contexts: sharing cookie dough logs, sharing fruit bars, sharing sub sandwiches and sharing cakes of different sizes.

Assessment Task Groups

List of Assessment Task Groups

A9.1:   Equally Sharing 3 Cookie Dough Logs among 4 Bakers

A9.2:   Equally Sharing 6 Fruit Bars among 5 Friends

A9.3: Equally Sharing 5 Sub Sandwiches among 7 People

A9.4: Equally Sharing 3 Different Cakes among 5 People

Task Group A9.1

Equally Sharing 3 Cookie Dough Logs among 4 Bakers

Materials: Multiple copies of a piece of paper with three identical rectangles on it, not aligned. For an example, see Figure 9.8.

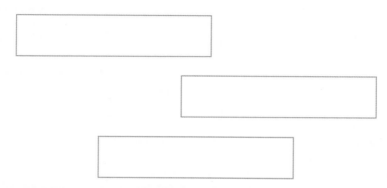

Figure 9.8 Sample organization of three cookie dough logs

What to do and say: Tell students this story: 'At a bakery the bakers make long cookie dough logs (make sure students know what this means) in preparation for slicing them and making cookies. This morning there are three long cookie dough logs made and four bakers to work with them. Show how to share the logs equally with the four bakers so that each one can get to work.' After students have marked the logs to make the equal shares, ask these follow-up questions:

1) Can you draw out the share for one baker separately?
2) How much does one baker get? Can you give a fraction name for this amount?
3) What is the size of one baker's share, compared to one log? How do you know?
4) What is the size of one baker's share, compared to all of the logs? How do you know?

Notes

- The reason to ask (1) is so that students can identify the share in relation to different quantities (one log, all of the cookie dough logs). However, Stage 1 students, who have not yet constructed a disembedding operation, may not find it a sensible request (see Chapters 4 and 5).

(Continued)

(Continued)

- The reason to ask (2) is to give students a chance to show how they initially view the share. Stage 2 students often have a dominant view of the share as a fraction in relation to all of the material but not in relation to one log, or vice versa. So, there is no need to specify what students are naming the amount in relation to, at least at first. The second question is listed here because students may not give a fraction name as a response to the first question.

- The reason to ask (3) is to specify that students compare the share (which they have drawn out separately) in relation to one log. If they don't find this sensible, you can rephrase as follows: 'Let's say one of the bakers says to you, "Hey, do I get a whole log or more than a log or less than a log?"' This rephrasing can help students focus on comparing the share in relation to one log. For those at the fifth level of fragmenting, a possible justification for (3) is, 'I know it's $\frac{3}{4}$ of one log because it's made of three one-fourths – three pieces that are each $\frac{1}{4}$ of one log.' Students who use repeated halving may give responses like 'one-half and one-fourth' or 'one-half and a smaller part'.

- The reason to ask (4) is to specify that students compare the share (which they have drawn out separately) in relation to all of the logs. For those at the fifth level of fragmenting, a possible justification for (4) is, 'I know it's $\frac{1}{4}$ because I was sharing all of the logs among four people' or 'I know it's $\frac{1}{4}$ because if I repeat the share 4 times I get 12 parts, which is all of the logs'. Students who are not yet at the fifth level of fragmenting may have a difficult time justifying their response to (4).

- A key point is that students who can respond to both (3) and (4), with justification, are likely at the fifth level of fragmenting. To confirm, it is best to follow up with Task Group A9.3 because sharing 5 items among 7 people is a more difficult problem than Task Group A9.1.

- Some students at the fifth level of fragmenting may not show distributive reasoning in this problem in the way Karenna demonstrated in this chapter. One reason is that the multiplicative relationship of $3 \times 4 = 12$ is quite familiar to some students, and so they may not have a need to engage in distributive reasoning (Lee and Aydeniz, 2015). This is another reason it is important to use Task Group A9.3, in which the multiplicative relationships are less familiar.

- Other number combinations could be used in this problem, to allow for using repeated halving where the share is smaller than one log: 2 logs shared among 4 bakers, 2 logs shared among 8 bakers, 3 logs shared among 8 bakers, 4 logs shared among 8 bakers, 5 logs shared among 8 bakers, 6 logs shared among 8 bakers, 7 logs shared among 8 bakers.

Task Group A9.2

Equally Sharing 6 Fruit Bars among 5 Friends

Materials: Multiple copies of a piece of paper with six identical rectangles on it, not aligned (similar to Figure 9.8).

What to do and say: Tell students that 5 friends want to share these 6 identical fruit bars. Ask them to make the equal shares. Then:

1) Can you draw out the share for one friend separately?
2) How much does one friend get? Can you give a fraction name for this amount?
3) What is the size of one friend's share, compared to one bar? How do you know?
4) What is the size of one friend's share, compared to all of the bars? How do you know?

Notes

- The first five notes in A9.1 apply here.
- This problem can be solved by sharing whole bars and then sharing just the sixth bar equally among the 5 friends. Other number combinations that could be used toward this end are: 4 bars shared among 3 people, 5 bars shared among 4 people, 7 bars shared among 6 people, 8 bars shared among 7 people.

Task Group A9.3

Equally Sharing 5 Sub Sandwiches among 7 People

Materials: Multiple copies of a piece of paper with five identical rectangles on it, not aligned (similar to Figure 9.8).

What to do and say: Tell students that 7 friends want to share these 5 identical sub sandwiches. Ask them to make the equal shares. Then:

1) Can you draw out the share for one person separately?
2) How much does one person get? Can you give a fraction name for this amount?
3) What is the size of one person's share, compared to one sandwich? How do you know?
4) What is the size of one person's share, compared to all of the sandwiches? How do you know?

Notes

- The first five notes in A9.1 apply here; please see also the discussion of Karenna's solution and the bricolage solution to this problem in the chapter. The solution via multiplication and division is also something students may demonstrate.
- This problem involves numbers that are relatively prime and so it may encourage or reveal distributive reasoning. Other number combinations that could be used toward this end are: 2 sandwiches shared among 3 people, 2 sandwiches shared among 5 people, 2 sandwiches shared among 7 people, 3 sandwiches shared among 5 people, 3 sandwiches shared among 7 people, 4 sandwiches shared among 7 people, 4 sandwiches shared among 9 people, 5 sandwiches shared among 9 people.

(Continued)

(Continued)

Task Group A9.4

Equally Sharing 3 Different Cakes among 5 people

Materials: Multiple copies of a piece of paper with three different-sized rectangles on it, not aligned (see Figure 9.9).

Figure 9.9   Three different-sized cakes

What to do and say: Tell students that 5 friends want to share these 3 different cakes equally. Ask them to make the equal shares. Then:

1) Can you draw out the share for one person separately?
2) What is the size of one person's share, compared to all of the cakes? How do you know?

Notes

- This problem can be used to further assess whether a student thinks of making equal shares for all three cakes by making that number of shares in each cake. However, in contrast with prior problems, in this problem the share is not three-fifths compared with one cake. It is only one-fifth compared with all of the cake.
- Students who justify (2) by repeating the share 5 times to make all of the cake demonstrate a key idea that is part of the fifth level of fragmenting.

Instructional Activities

List of Instructional Activities

IA9.1: Equally Sharing Multiple Items of the Same Size

IA9.2: Equally Sharing Multiple Items of Different Sizes

Activity IA9.1

Equally Sharing Multiple Items of the Same Size

Intended learning: This activity is intended to support students' initial development of distributive reasoning.

Instructional mode: Game played in pairs.

Materials: Unmarked fraction strips of the same size, pencils, scissors, tape.

Description: Using the bakery context introduced in Task Group A9.1, the teacher should give each pair of students several fraction strips, telling them that these strips represent cookie dough logs. One student in the pair gets to decide the number of logs to be shared, and the other student gets to decide how many bakers there will be. Then the students work together to determine how much of the cookie dough logs each baker should receive.

Responses, Variations and Extensions

- Students can use any of the strategies mentioned in this chapter – repeated halving, sharing wholes first, using bricolage, or distributive reasoning – to solve these problems. Some choices of numbers will lend themselves better to one strategy or another. For example, if the number of bakers is a power of 2 (2, 4, or 8), students will be able to successfully apply the repeated halving strategy.
- As such, solving these problems can support refining computational strategies.
- Solving the problems can help students develop understanding that sharing each of the logs equally yields the same solution as sharing all of the logs equally – distributive reasoning.
- These problems can be posed in other contexts, such as the fruit bars or sandwiches contexts.
- Problems like these can support standards related to understanding fractions as quotients – standards like Common Core State Standard 5.NF.3. At the same time, they complexify arithmetic by extending whole number division to problems that do not have whole number solutions.

(Continued)

(Continued)

Activity IA9.2

Equally Sharing Multiple Items of Different Sizes

Intended learning: This activity is intended to solidify students' development of distributive reasoning.

Instructional mode: Whole-class activity.

Materials: Copies of Figure 9.9, pencils, scissors, tape.

Description: The teacher should give each student a copy of Figure 9.9 (three differently sized rectangles) and tell the students that the rectangles represent cakes of three different sizes. Then, the teacher should ask the students to make $\frac{1}{7}$ of each cake. Students should notice that the three $\frac{1}{7}$ pieces are different sizes, and they should be able to justify this on the basis that the wholes were different sizes. Next, the teacher should ask them to make $\frac{1}{7}$ of the three cakes, together: 'Show me how much you would get if you were given one-seventh of ALL of the cake – all three cakes together.' Then the teacher should ask the students to justify how they know they have produced $\frac{1}{7}$ of all the cake.

Responses, Variations and Extensions

- Like the problems in Activity IA9.1, this problem can be posed in other contexts (fruit bars, sandwiches, etc.) and using different numbers of items or fractions. For example, asking for $\frac{3}{7}$ of all the cake is a harder question than $\frac{1}{7}$ of all the cake.
- Problems like this one, where the items are different sizes, are more challenging because the shares are distributed over differently sized wholes, producing differently sized unit fractions.

10

Teaching Students at Stages 2 and 3: Multiplying Fractions

Students at Stages 2 and 3 can learn to engage in a wide range of reasoning with fractions, including reasoning to develop arithmetical operations with fractions (Hackenberg and Tillema, 2009; Steffe and Olive, 2010). However, the learning pathways of the students at different stages are different. For example, students at both stages can develop conceptual meaning for taking fractions of fractions (multiplying fractions), but students at Stage 2 will be unlikely to develop as general a way of thinking as students at Stage 3. This comment holds true for combining fractions additively as well as dividing fractions.

This chapter focuses on how students at Stages 2 and 3 can learn to multiply fractions based on reasoning. In Chapter 11 we address how students at these two stages can learn to add and subtract fractions, and in Chapter 12 we focus on how they can learn to divide fractions. In each of these chapters we address characteristics of students at each stage, challenges they may encounter at each stage and ways to support students in working on those challenges.

Taking Unit Fractions of Unit Fractions

In prior chapters (4, 5, 6 and 9) we addressed equal sharing of single and multiple items as an important basis for developing meaning for fractions. In our view, sharing equal shares is a way into fraction multiplication. For example, consider this problem:

'Latecomer Problem' You share the rectangular loaf of cranberry bread shown between yourself and four friends (Figure 10.1), so five people in total share the bread.

1)   Make the shares and draw out your piece separately.

2)   Two friends come late, and you agree to share your piece equally with them, so three people share your piece. Make the piece of bread you get now. Draw it out separately.

3)   How much of the whole bread do you get now? How do you know? Explain.

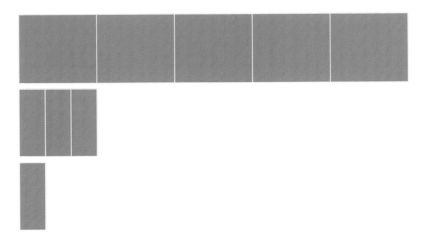

Figure 10.1   Work on the 'Latecomer Problem'

From an expert adult perspective, this problem is about taking $\frac{1}{3}$ of $\frac{1}{5}$ of a piece of bread. Students who have worked on equal sharing problems and fraction language may also see it that way, and a common initial answer for some students is: 'I get one-third.' They are correct! But the question is, 'One-third of what?' The questions in part (3) can help students to articulate that the piece is much smaller than one-third of the whole loaf of bread.

Students at both Stages 2 and 3 are likely to see that they could partition each of the five parts of the bread into three equal parts and that would make 15 equal parts in all (Hackenberg and Tillema, 2009). So that means their new piece is $\frac{1}{15}$ of the bread because it is one of those 15 equal parts. Asking students to articulate how they determined the 15 equal parts is helpful because it is an example of how whole number multiplication is relevant in taking fractions of fractions, which eventually students can learn to see as multiplying fractions. That is, students at Stages 2 and 3 can and should articulate how $5 \times 3 = 15$ allowed them to solve the problem.

Here are two problems closely related to the 'Latecomer Problem', but with important differences:

'Gum Tape Problem': Phoebe got a piece of gum tape, 1 metre long, for her birthday (Figure 10.2). She has $\frac{1}{3}$ of it left. Her mother says she can have $\frac{1}{5}$ of that amount to chew right now.

1) To start with, draw the amount of tape that Phoebe has left. Draw it as a separate amount from the 1-metre-long tape. Don't convert to centimetres! Stay in metres.

2) Draw the amount of tape Phoebe's mother says she can chew right now. Draw it as a separate amount from the 1-metre-long tape and your drawing in part (1).

3) How much of a metre does Phoebe get to chew right now? Don't convert to centimetres! Stay in metres.

4) Explain your solution.

Figure 10.2 Rectangle representing 1 metre of gum tape

'Unit Fractions of Unit Fractions Problem': Imagine a long rectangular sub sandwich. You get $\frac{1}{2}$ of $\frac{1}{6}$ of it. Can you imagine taking $\frac{1}{2}$ of $\frac{1}{6}$ of that rectangular sub sandwich? [The same problem can be done with many different unit fractions.]

a) Explain to your partner how you are imagining this problem.

b) How much of the whole sandwich do you get to eat? How do you know?

c) Draw what you imagined and justify your answer to (2). [The drawing part can be omitted if the whole class is working just on mentally visualizing.]

The 'Gum Tape Problem' uses fraction language rather than equal sharing language, so it is more advanced in that students have to activate partitioning operations to solve the problem. If students do not activate partitioning operations to solve problems like this one, we suggest returning to sharing language and using fraction language like 'one of three equal parts', as we discuss in Chapter 4. When students are transitioning to or have transitioned to fraction language, the teacher can notate reasoning on the 'Gum Tape Problem' as follows:

$\frac{1}{5}$ of $\frac{1}{3}$ = ?

$3 \times 5 = 15$ equal parts in the whole metre

So, $\frac{1}{5}$ of $\frac{1}{3}$ = $\frac{1}{15}$ of a metre

We emphasize that this notation can help students link whole number multiplication to fraction multiplication, but it showcases reasoning as opposed to the very compressed notation of the standard algorithm. In our view, it is important to hold off on that level of compression for some time.

We also point out that we have used 'of' instead of a multiplication symbol in ' $\frac{1}{5}$ of $\frac{1}{3}$ '. The main reason is that the idea of taking $\frac{1}{5}$ of $\frac{1}{3}$ is often sensible to students, whereas using the multiplication symbol to represent this idea often is not (Behr et al., 1993; Hackenberg, 2010). So, we suggest using the natural language, 'of,' for some time. We understand that teachers often tell students that in mathematics 'of' means multiply, but we recommend refraining from that here until students have had a lot of experience taking fractions of fractions.

Also in contrast with the 'Latecomer Problem', the 'Gum Tape Problem' involves taking $\frac{1}{5}$ of $\frac{1}{3}$, instead of $\frac{1}{3}$ of $\frac{1}{5}$. In these two problems, the process of picture drawing is different, but the result is the same. Discussing this issue can provide an important foundation for similar, but more complex, discussions as the work on taking fractions of fractions continues.

A next step after working on problems like the 'Gum Tape Problem' is to visualize taking unit fractions of unit fractions mentally, without drawing (thus **distancing the instructional setting**). The 'Unit Fractions of Unit Fractions Problem' asks students to imagine taking a unit fraction of a unit fraction. When students can think this way in their imagination, then they do not have a need to memorize a rule about multiplying denominators. Furthermore, these problems can promote making generalizations: developing efficient methods that are based on reasoning but do not require students to go through the whole process of reasoning. For example, after solving problems mentally, students might make a generalization like, 'Oh, I multiply the denominators in the two fractions because it finds the total number of equal parts in my whole. Once I know that, I know the fraction size of the parts.'

Taking Non-Unit Fractions of Unit Fractions

Once students have developed ways of taking unit fractions of unit fractions, a good next step is to take non-unit fractions of unit fractions. To be accessible to students at both Stages 2 and 3, it is best to start with non-unit proper fractions. For example, consider this problem:

'Ribbon Problem': Janelle has $\frac{1}{5}$ of a yard of ribbon. She uses $\frac{2}{3}$ of that amount to make a bracelet.

1) To start with, make a drawing of a yard of ribbon.

2) Draw the amount of ribbon that Janelle has. Draw it as a separate amount.

3) Draw the amount of ribbon that Janelle uses. Draw it as a separate amount.

4) How much of a yard does Janelle use? Don't convert to inches! Stay in yards.

5) Explain your solution.

This problem involves taking a proper fraction ($\frac{2}{3}$) of a unit fraction ($\frac{1}{5}$). A problem like this one is a good next step because taking multiple parts from a single part is not much harder than taking a single part from a single part, as in the 'Gum Tape Problem'. That is, students who can take $\frac{1}{3}$ of $\frac{1}{5}$ will not find it much harder to take $\frac{2}{3}$ of $\frac{1}{5}$, because they will take two parts and have developed a way of thinking to know that each of those parts is $\frac{1}{15}$ of the whole unit, a yard in this case. Still, this new problem helps students to **extend the range of numbers** that they can multiply.

A teacher can notate reasoning on the 'Ribbon Problem' as follows:

$\frac{2}{3}$ of $\frac{1}{5}$ = 2 × ($\frac{1}{3}$ of $\frac{1}{5}$)

5 × 3 = 15 equal parts in the whole yard

So, each part is $\frac{1}{15}$ of a yard, so 2 × $\frac{1}{15}$ = $\frac{2}{15}$ of a yard

Following problems like the 'Ribbon Problem', it is good to pose problems that encourage students to visualize without drawing, analogous to the 'Unit Fractions of Unit Fractions Problem'.

Taking Unit Fractions of Non-Unit Fractions

A more difficult step is to take a unit fraction of a non-unit fraction. For example, consider this problem:

'Liquorice String Problem': Mio had $\frac{2}{3}$ of a yard of liquorice string. He ate $\frac{1}{5}$ of that amount.

1) To start with, make a drawing of a yard of liquorice.

2) Draw the amount of liquorice string that Mio had initially. Draw it as a separate amount.

3) Draw the amount of liquorice string that Mio ate, but don't erase the mark between your two $\frac{1}{3}$ yards. Draw what Mio ate as a separate amount.

4) How much of a yard did Mio eat? Don't convert to inches! Stay in yards. If you can't determine the amount from your drawing, then try part (3) again.

5) Explain your solution.

This problem involves taking a unit fraction ($\frac{1}{5}$) of a proper fraction ($\frac{2}{3}$), which means taking a single part from multiple parts. To do so is challenging for both Stage 2 and Stage 3 students, but students operating at Stage 3 are likely to be able to address the challenge more efficiently and effectively than students at Stage 2 (Hackenberg and Tillema, 2009).

A key way to address the challenge is to realize that to take $\frac{1}{5}$ of any quantity means taking $\frac{1}{5}$ of each part of that quantity. The root for this way of thinking is what we discussed in Chapter 9: taking $\frac{1}{5}$ of two bars can be accomplished by taking $\frac{1}{5}$ of each bar. We described how this way of thinking is distributive reasoning – that is, it reflects the Distributive Property of Multiplication over Addition. A way of **notating** this reasoning in the 'Liquorice String Problem' is as follows:

$\frac{1}{5}$ of $\frac{2}{3}$ = $\frac{1}{5}$ of ($\frac{1}{3}$ + $\frac{1}{3}$) = $\frac{1}{5}$ of $\frac{1}{3}$ + $\frac{1}{5}$ of $\frac{1}{3}$ = $\frac{1}{15}$ + $\frac{1}{15}$ = $\frac{2}{15}$

We are not suggesting that all students who solve this problem will be ready to write notation in this way. But we highlight it for three reasons. First, it is consistent with our general recommendation for developing notation that represents reasoning (**notating**). Second, it can highlight ideas for students that otherwise could remain implicit, such as that they are thinking of $\frac{2}{3}$ as $\frac{1}{3} + \frac{1}{3}$. Third, it can help teachers see that a student who is reasoning this way is using the Distributive Property.

Taking Fractions of Fractions, Generally

In our experience, only students at Stage 3 are poised to develop a general way of thinking for taking any fraction of any fraction (Hackenberg and Tillema, 2009; Steffe and Olive, 2010). The main reason is that doing so requires coordinating multiple embedded units. We illustrate with an example:

'Pastry Chef Problem': Oliver, a pastry chef, had $\frac{4}{5}$ of a kilogramme of sugar on his counter. He used $\frac{2}{3}$ of that amount in a recipe. How much of a kilogramme did he use in the recipe?

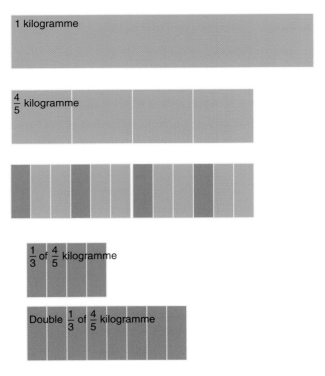

Figure 10.3   Taking $\frac{1}{3}$ of $\frac{4}{5}$ of a kilogramme and then doubling it to get $\frac{2}{3}$ of $\frac{4}{5}$ of a kilogramme

1) To start with, draw a bar to represent one kilogramme of sugar.

2) Draw the amount of sugar that Oliver had. Draw it as a separate amount.

3) Draw the amount of sugar that Oliver used in the recipe. Don't erase the marks between your four $\frac{1}{5}$ s. Draw what Oliver used as a separate amount.

4) How much of a kilogramme does Oliver use? If you can't determine it from your drawing, then try part (3) again.

5) Explain your solution.

There are at least two main ways to solve this problem using distributive reasoning. One way is to take $\frac{1}{3}$ of $\frac{4}{5}$ of a kilogramme by taking $\frac{1}{3}$ of each of the four $\frac{1}{5}$-kilogrammes. Then a student can double that to get $\frac{2}{3}$ of $\frac{4}{5}$ of a kilogramme (Figure 10.3).

Another way is to take $\frac{2}{3}$ of each of the four $\frac{1}{5}$-kilogrammes, which means getting $\frac{2}{15}$ of a kilogramme 4 times (Figure 10.4). We find this second way to be more common among students, but that may vary quite a bit among classes.

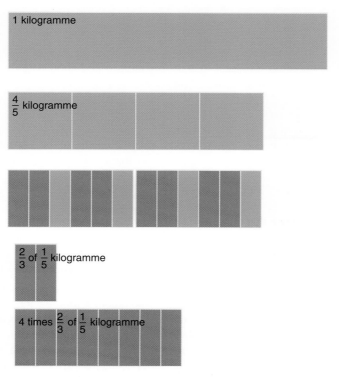

Figure 10.4 Taking $\frac{2}{3}$ of each $\frac{1}{5}$ of a kilogramme, or $\frac{2}{15}$ of a kilogramme 4 times

We have already noted that students at Stage 2 find the distributive reasoning involved in either solution to be challenging. So, for that reason alone solving this problem can be quite difficult.

In addition, let's say that a student at Stage 2 has taken $\frac{2}{3}$ of each $\frac{1}{5}$-kilogramme and drawn out those eight parts. These students are quite likely to name the result $\frac{8}{12}$ of a kilogramme (Hackenberg and Tillema, 2009). Why? Because naming the result as $\frac{8}{15}$ of a kilogramme requires keeping track of $\frac{4}{5}$ of a kilogramme as a unit of 4 units in relation to the whole kilogramme as a unit of 5 units. Students at Stage 2 can keep track of only two levels of units at a time as they continue to engage in mathematical activity. We have found that, when questioned, some Stage 2 students will agree that the amount should be $\frac{8}{15}$ of a kilogramme, not $\frac{8}{12}$ of a kilogramme. But it is quite likely they will repeatedly make this kind of conflation in their reasoning.

In contrast, students at Stage 3 can keep track of the multiple embedded units. They are likely to find it helpful to notate their reasoning as follows (in accord with the solution shown in Figure 10.4):

$$\frac{2}{3} \text{ of } \frac{4}{5} = \frac{2}{3} \text{ of } (\frac{1}{5} + \frac{1}{5} + \frac{1}{5} + \frac{1}{5})$$

$$= \frac{2}{3} \text{ of } \frac{1}{5} + \frac{2}{3} \text{ of } \frac{1}{5} + \frac{2}{3} \text{ of } \frac{1}{5} + \frac{2}{3} \text{ of } \frac{1}{5}$$

$$= \frac{2}{15} + \frac{2}{15} + \frac{2}{15} + \frac{2}{15}$$

Furthermore, they may find it helpful to start to compress and abbreviate their reasoning by recognizing that they are finding $\frac{2}{3}$ of $\frac{1}{5}$ four times:

$$\frac{2}{3} \text{ of } \frac{4}{5} = (\frac{2}{3} \text{ of } \frac{1}{5}) \times 4$$

As we noted in Chapter 9, this notation reflects *associative reasoning,* or the Associative Property of Multiplication. A teacher can work with students to see that this could also be broken apart to:

$$\frac{2}{3} \text{ of } \frac{4}{5} = (\frac{2}{3} \text{ of } \frac{1}{5}) \times 4 = 2 \times (\frac{1}{3} \text{ of } \frac{1}{5}) \times 4$$

This notation helps to reveal that the core of taking any fraction of any fraction is finding the size of the parts by thinking about the unit fraction of unit fraction embedded in the problem. In this case, we are taking thirds of fifths of a kilogramme, which produces fifteenths of a kilogramme. Then the issue is to determine how many of those fifteenths are needed. Since Oliver wants $\frac{2}{3}$ of the 4 equal fifths, he needs to take 2 parts from one-fifth 4 times, or 8 parts total.

This kind of thinking can lead to developing a sound rationale for the standard computational algorithm for fraction multiplication (Tillema and Hackenberg, 2011). Indeed, this notation shows that to find $\frac{2}{3}$ of $\frac{4}{5}$ one can multiply numerators and multiply denominators. However, we caution against reducing student reasoning to this rule for quite some time. Students need a great deal of repeated experience with reasoning about taking fractions of fractions in the way we have described here in order to develop generalizations such as 'multiply numerators and multiply denominators', where the generalizations point to the abbreviation of reasoning.

Assessment Task Groups

List of Assessment Task Groups

A10.1:   Sharing Shares

A10.2:   Taking Fractions of Unit Fractions

A10.3:   Taking Unit Fractions of Fractions

A10.4:   Taking Fractions of Fractions

Task Group A10.1

Sharing Shares

Materials: Unmarked fraction strips (see Appendix), scissors, writing utensils.

What to do and say: Tell students the fraction strip represents a loaf of banana bread: 'You and five friends are going to share this loaf as an after-school treat, so six people are sharing the bread. Show how to make the equal shares, and draw or cut out your piece. Then two neighbours come late, and you agree to share your piece with them. Show how to make those shares and draw or cut out the piece you get now. What fraction of the whole loaf do you get now? How do you know?'

Notes

- Students at Stage 1 may persist in calling the amount of bread they get 'one-third' (Hackenberg and Tillema, 2009). If students persist in this way, ask them how many parts of that size they need to make the whole cake. Students at Stage 1 may be able to determine a solution and name the part 'one-eighteenth'. They will sometimes make this determination by repeating the part until they make the whole bread, rather than by thinking of partitioning each of the six parts into three equal parts. Even if they do partition each of the six parts into three equal parts, they may not draw upon this activity in solving future similar problems. The main reason is that the sharing shares questions above may not initiate for them the ideas of inserting parts inside parts; only more direct questions trigger that.
- For some students, you may want to use fewer friends (three or four) and have just one neighbour come late, in order to find out if the students can make progress with halving a smaller number of parts.
- For some students, you might follow up by asking how much the two neighbours get ($\frac{2}{18}$). If they say $\frac{2}{18}$, then that is evidence that they see each part as $\frac{1}{18}$ and they can iterate that twice. So, this solution is evidence that students have developed a Measurement Scheme for Unit Fractions, as discussed in Chapter 6.

(Continued)

(Continued)

- You could also ask how much a friend gets now, and whether there are two names for a friend's amount. If students say a friend gets $\frac{1}{6}$ and also $\frac{3}{18}$, ask them to tell you why the friend's amount has these two names. Students may articulate that the friend still gets $\frac{1}{6}$ because they have not shared their piece with anyone else. But three eighteenths fit into $\frac{1}{6}$, so a friend's piece can also be named $\frac{3}{18}$. We discuss different names for the same-size fraction (commensurate fractions) in Chapter 11.

Task Group A10.2

Taking Fractions of Unit Fractions

Materials: Unmarked fraction strips, scissors, writing utensils.

What to do and say: Tell students the fraction strip represents a long piece of ribbon. At a costume store, workers cut off the amount of ribbon needed to decorate costumes and masks. Wanda has used some of the ribbon shown and there is now $\frac{1}{3}$ of it left. Draw how much is left as a separate piece of ribbon. Jamal needs $\frac{1}{4}$ of that amount. Draw how much Jamal needs as a separate piece of ribbon. How much of the whole ribbon did Jamal use? How do you know?

Notes

- For variation and to find out how students do in taking non-unit fractions of $\frac{1}{3}$, tell them Jamal or another worker now needs $\frac{3}{4}$ of the $\frac{1}{3}$, or $\frac{2}{5}$ of the $\frac{1}{3}$.
- For some students, you could say that Jamal or another worker now needs $\frac{7}{5}$ of the $\frac{1}{3}$. Doing so could allow you to further test out whether they have developed fractions as numbers (see Chapter 8).

Task Group A10.3

Taking Unit Fractions of Fractions

Materials: Multiple copies of a piece of thick paper (ideally cardstock) with one rectangle on it and with $\frac{2}{3}$ of that rectangle shown. The mark separating the $\frac{2}{3}$-rectangle into thirds should be present (see Figure 10.5).

1 slab

Figure 10.5   Set-up for Taking Unit Fractions of Fractions

Also needed are scissors and a cut-out bar that shows an unmarked $\frac{2}{3}$-bar partitioned into fifths, with one of those parts cut out as well (Figure 10.6).

Figure 10.6 $\frac{1}{5}$ of $\frac{2}{3}$ with the mark cleared

What to do and say: Tell students the original rectangle represents a long rectangular slab of clay. At a pottery studio, artists are taking parts of the clay to create their pottery. The store owner, Sarina, keeps track of what they use. Right now Sarina has $\frac{2}{3}$ of this slab left for use, as shown on the paper. Pedro, the potter, comes up and asks for $\frac{1}{5}$ of that amount. Draw how much Pedro needs as a separate slab. However, at the end of drawing, students must know what fraction of 1 slab he is using so Sarina can record it. To accomplish this, advise students not to erase or ignore the mark that separates the two $\frac{1}{3}$s on the $\frac{2}{3}$-bar.

Notes

- This is a tough question, and it is not necessary for students to complete it to get an understanding of their ideas about it.
- Some students will partition the $\frac{2}{3}$-slab into five equal parts, ignoring the thirds mark, and draw out one of those parts. However, they will not know how much that part is of the whole slab. You can ask them how they could figure it out, and they might indicate they want to see how many times that part fits into the whole slab. You can suggest they use scissors to work on this, assisting in cutting out their part as needed. If they iterate the part along the slab, they should see that the part will not fit into the slab a whole number of times. If they think that is because their marks were not perfect, you can provide a cut-out part (Figure 10.6) that was made 'perfectly'. In iterating it, they will still see that the part will not fit into the slab a whole number of times.
- Overall, encourage students not to erase or ignore the thirds mark. If they start down a solution path where they do that, steer them back to not erasing the thirds mark. If they are stumped, you can praise them for trying and tell them you do not expect them to know all of these problems; you will come back to work on this later.
- If students are stumped here, it is not necessary to work on A10.4.

(Continued)

(Continued)

Task Group A10.4

Taking Fractions of Fractions

Materials: Multiple copies of a piece of thick paper (ideally cardstock) with one rectangle on it

What to do and say: This is the same set-up as in A10.3, except that another potter, Maya, wants $\frac{3}{5}$ of the $\frac{2}{3}$-slab. Again, at the end of their drawing, students must know what fraction of 1 slab she is using so Sarina can record it. To accomplish this, tell students not to erase the thirds mark.

Notes:

- For some students you may want to test out using an improper fraction. For example, Sarina might have $\frac{4}{3}$ of a slab, and Maya wants $\frac{3}{5}$ of that amount. If students understand $\frac{4}{3}$ as a number (see Chapter 8), this problem could be a good challenge.
- Alternately, Sarina might have $\frac{2}{3}$ of a slab and Maya wants $\frac{7}{5}$ of that amount. This problem is even more challenging because Maya wants more than the material available. If students are concerned about that, you can say that Sarina can add in more clay if Maya needs it, similar to the idea of magic candy bars in Chapter 8. An even greater challenge is for Sarina to have $\frac{4}{3}$ of a slab and for Maya to want $\frac{7}{5}$ of that amount.

Instructional Activities

List of Instructional Activities

IA10.1:   Sharing Shares

IA10.2:   Taking Fractions of Fractions I

IA10.3:   Making $\frac{1}{4}$ of a Recipe

IA10.4:   Taking Fractions of Fractions II

IA10.5:   Notating and Articulating Patterns in Reasoning

Activity IA10.1

Sharing Shares

Intended learning: Students will learn to share shares and name the result in relation to the designated unit.

Instructional mode: Students working in pairs or small groups with instruction from the teacher.

Materials: Unmarked fraction strips (see Appendix), scissors, writing utensils, dice.

Description: Each pair or group will solve a version of the 'Latecomer Problem'. To determine the numbers of friends and neighbours for each pair or group, the pair/group will roll a pair of dice, with one die representing the number of friends and the other die representing the number of neighbours. Students are to show the process of making the shares and how they determined what the size of their share is in relation to the whole loaf of bread. Students can present and explain their solutions to the class.

Responses, Variations and Extensions

- Students can be asked about the size of the share for multiple neighbours or the size of a share for a friend, as discussed in this chapter.
- If students are not yet describing their process with fraction language, teachers can help them to articulate it using the 'parts of parts' language, as developed in Chapter 4. For example, 'I took one of three equal parts from one of five equal parts.' This language can get cumbersome when problems increase in complexity, but it may be appropriate for some time with sharing shares.
- If students are using fraction language to describe their process, teachers can involve them in notating their reasoning here as outlined for the 'Gum Tape Problem'.
- This activity could be done with *JavaBars* (math.coe.uga.edu/olive/welcome.html).
- This activity could be played as a game, where students earn points for correct solutions and correct explanations of their solutions.
- Instead of dice, pairs of students could use a pack of regular playing cards to facilitate the game. Each player draws a card, and the number on the card would represent the number of equal shares to make. One player could be the initial sharer and the other could be the 'second sharer' (the sharer of one share from the first round of sharing). Or, the pair could figure out the sharing of shares in both ways. Initially students could draw out the equal shares, and later they could imagine them and then discuss what they imagined and their results. Replacing dice with cards is a means of extending the range of numbers (from 1–6, to 1–13).

Activity IA10.2

Taking Fractions of Fractions I

Intended learning: Students will learn to take unit fractions and non-unit proper fractions of a unit fraction; they will also learn to imagine this process and to write notation to represent it.

Instructional mode: Students working in pairs or small groups with instruction from the teacher, followed by whole classroom instruction and discussion.

Materials: Unmarked fractions strips, writing utensils.

Description: Pose the 'Gum Tape Problem' to each pair or group with a different pair of unit fractions for each group. It can be advantageous to include the same fractions but in reverse order. For example, one group could take $\frac{1}{3}$ of $\frac{1}{5}$ of the metre, and another group could take $\frac{1}{5}$ of $\frac{1}{3}$ of the metre. After class presentations of solutions and further problems in which students draw their solutions, engage the whole class in the 'Unit Fractions of Unit Fractions Problem', where students imagine the process of taking a unit fraction of a unit fraction. Discuss what students are imagining and write notation, as shown for the 'Gum Tape Problem' to represent their solution processes.

(Continued)

(Continued)

Responses, Variations and Extensions

- This activity could be done with *JavaBars*.
- As students become adept at drawing and visualizing problems in which they take a unit fraction of a unit fraction, you can ask them to take a non-unit fraction of a unit fraction. This is a means of **complexifying** the arithmetic.
- This activity could be played as a game, where students earn points for correct solutions and explanations of their solutions that are grounded in making the fraction parts. The process would be similar to the game variation in IA10.1, except we recommend making a set of cards with unit fractions on them. As students make progress, the teacher could include some non-unit fraction cards, where the non-unit fractions are to be taken of unit fractions.

Activity IA10.3

Making $\frac{1}{4}$ of a Recipe

Intended learning: Students will learn to take a unit fraction of a non-unit fraction and notate their reasoning.

Instructional mode: Students working individually or in pairs with instruction from the teacher, followed by whole classroom instruction.

Materials: Paper with a rectangle on it to represent 1 kilogramme of flour, with $\frac{3}{5}$ kilogramme represented underneath. The marks between the three one-fifths should be well visible (see Figure 10.7).

1 kilogramme of flour

$\frac{3}{5}$ kilogramme of flour

Figure 10.7 Set-up for the 'Muffin Problem'

Description: Pose the 'Muffin Problem' to all students. Tell them: A honey wheat muffin recipe requires $\frac{3}{5}$ kilogramme of whole-wheat flour. However, Nikolas wants to make $\frac{1}{4}$ of the recipe. How much flour does he need? The rectangle on your paper represents 1 kilogramme of flour, and the rectangle below it represents $\frac{3}{5}$ kilogramme of flour. Draw what Nikolas needs, but don't erase or ignore the marks separating the three $\frac{1}{4}$ kilogrammes. Based on your drawing, determine how much of a kilogramme of flour Nikolas needs. Explain your reasoning.

Pose the 'Muffin Problem' (or the 'Liquorice String Problem' from this chapter) with other fractions, where students are always taking a unit fraction of a non-unit fraction. Here are some good numbers to use: $\frac{1}{3}$ of $\frac{4}{5}$ kg, $\frac{1}{5}$ of $\frac{2}{3}$ kg, $\frac{1}{5}$ of $\frac{3}{4}$ kg. Initially you could supply the start of the drawing for students as above. Later you could ask students to make their own drawings. After they have had some repeated experiences with solving these problems, lead a discussion about what is in common across their solution and develop notation for their reasoning. We have outlined how to notate distributive reasoning in this chapter. We discuss another common solution process and notation of it below.

Responses, Variations and Extensions

- Some students may mark each of the three parts in the $\frac{3}{5}$-bar into four equal mini-parts. They will make 12 mini-parts in all, and they may determine that three mini-parts is $\frac{1}{4}$ of 12. Then they may call the result $\frac{3}{12}$. If so, engage students in a discussion about what the $\frac{3}{12}$ refers to: where are the 12 equal mini-parts in the whole kilogramme? Students are likely to see that the 12 mini-parts are in just $\frac{3}{5}$ of the kilogram, and that there would be 20 of these mini-parts in the whole kilogram. So, the result is $\frac{3}{12}$ of $\frac{3}{5}$-kilogramme but $\frac{3}{20}$ of 1 kilogramme.

- In the above solution, students are probably not engaging in distributive reasoning. Instead they are making a whole number of parts that is divisible by four, since in this problem they are aiming to make $\frac{1}{4}$. This is a valid solution process, although it does not generalize as nicely as distributive reasoning does. However, we suggest trying to write notation to represent and honour the reasoning process. For example: $3 \times 4 = 12$ equal parts in the $\frac{3}{5}$-kg bar. $12 \div 4 = 3$, so 3 of those parts are $\frac{1}{4}$ of the $\frac{3}{5}$-kg. $5 \times 4 = 20$ equal parts in the 1 kg bar. So, those 3 parts are $\frac{3}{20}$ of 1 kg.

- You can start this problem with an amount of flour consisting of just two fraction parts (e.g. $\frac{2}{3}$ of $\frac{2}{5}$ kg), but then ask students to make $\frac{1}{3}$ of it.

- This activity could be done with *JavaBars*. This software is particularly useful because it facilitates the issue of not erasing the marks between parts on the initial fraction.

Activity IA10.4

Taking Fractions of Fractions II

Intended learning: Students will learn to take non-unit fractions of non-unit fractions.

Instructional mode: Students working individually or in pairs with instruction from the teacher.

Materials: Paper with a rectangle on it to represent one unit of the quantity in the problem (e.g. 1 kg of sugar), with the initial fraction amount of the quantity underneath. The marks between the fraction units should be well visible (see Figure 10.7).

(Continued)

(Continued)

Description: Pose the 'Pastry Chef Problem' or problems that are similar to it. Other contexts that could be used are tonnes of soil at a gardening shop, kilogrammes of clay at a pottery studio, kilogrammes of bulk items (grains, beans, nuts, dried fruit) at a grocery store, cups of spices at a grocery store. In general it is helpful to use weights and capacities because it is somewhat impractical to cut apart lengths, since it is usually not practical to recombine them again. However, you can use length contexts as we have in the 'Liquorice String Problem' if it seems believable to students to have separate parts (i.e. it is believable to eat a few separate parts of liquorice string).

Here are some fractions to use: $\frac{2}{3}$ of $\frac{2}{5}$, $\frac{2}{3}$ of $\frac{5}{7}$, $\frac{2}{5}$ of $\frac{3}{4}$, $\frac{2}{5}$ of $\frac{4}{9}$, $\frac{3}{4}$ of $\frac{5}{9}$, $\frac{3}{4}$ of $\frac{7}{8}$, $\frac{3}{5}$ of $\frac{4}{7}$, $\frac{3}{5}$ of $\frac{7}{8}$.

Responses, Variations and Extensions

- The same kind of responses as described in the first bullet of IA10.3 can occur when students solve the 'Pastry Chef Problem', as discussed in this chapter.
- To challenge students who are ready, use improper fractions. For example, ask them to take $\frac{2}{3}$ of $\frac{7}{5}$, $\frac{4}{3}$ of $\frac{2}{5}$, or $\frac{4}{3}$ of $\frac{11}{9}$. These problems are quite challenging because the units coordination load is increased (see Chapter 8).
- This activity could be done with *JavaBars*. This software is particularly useful because it facilitates the issue of not erasing the marks between parts on the initial fraction.
- As students move from unit fractions in earlier activities to non-unit fractions here, they are complexifying the arithmetic of multiplication.
- At the same time, students are addressing fractions multiplication standards like Common Core State Standard 5.NF.3.

Activity IA10.5

Notating and Articulating Patterns in Reasoning

Intended learning: Students will learn to notate and articulate patterns in their reasoning for taking non-unit fractions of non-unit fractions.

Instructional mode: Whole class instruction and discussion.

Materials: Student work from IA10.4.

Description: Tell students that we are going to review their work on taking fractions of fractions and describe patterns in their ways of thinking. Start with a problem that all students have solved, such as the 'Pastry Chef Problem', and write notation to represent the reasoning that they have done.

Responses, Variations and Extensions

- If students have worked on this in IA10.3, they might be able to lead some of this. Or, the teacher may lead.

- In the process of **notating**, ask students to articulate common ways of thinking. For example: 'I took the given fraction of each unit fraction part in the kilogrammes of flour' would be a way students might articulate distributive reasoning.

- To move from recording distributive reasoning into a more compressed form of notation (toward the standard computational algorithm – **formalizing**), the teacher can lead a discussion similar to the discussion in the last part of this chapter with the 'Pastry Chef Problem'. Following this discussion, we recommend giving students some notation and asking them to say what the problem situation could be that would lead to that. For example, $3 \times (\frac{1}{4} \times \frac{1}{5}) \times 2$ could be seen as taking $\frac{3}{4}$ of $\frac{2}{5}$ of a kg of flour in order to make $\frac{3}{4}$ of the recipe.

11

Teaching Students at Stages 2 and 3: Adding and Subtracting Fractions

Domain Overview

Many people believe that addition is a more basic operation than multiplication. However, some basic ideas about multiplying fractions are needed for adding and subtracting fractions. To see why this is so, consider this key idea when adding and subtracting fractions: fractions to be added or subtracted need to be measured in equal size parts. For example, $\frac{1}{2}$ of a unit and $\frac{1}{3}$ of that same size unit can be added by joining them (Figure 11.1).

However, if a person wants to name the result of $\frac{1}{2} + \frac{1}{3}$ with a single fraction, then they have to measure each amount in the same size units. To do so requires measuring $\frac{1}{2}$ and $\frac{1}{3}$ with the same size fraction, which involves thinking about how to measure each fraction with smaller fractions. For example, $\frac{1}{2}$ can be measured with two $\frac{1}{4}$s because $\frac{1}{2}$ of $\frac{1}{2}$ is $\frac{1}{4}$. Note the use of unit fraction multiplication here. One-third can be measured with two 1/6s because $\frac{1}{2}$ of $\frac{1}{3}$ is $\frac{1}{6}$, another use of unit fraction multiplication. However, it is not possible to measure $\frac{1}{3}$ with a whole number of $\frac{1}{4}$s, because two $\frac{1}{4}$s already make $\frac{1}{2}$. Can $\frac{1}{2}$ be measured with $\frac{1}{6}$s? Let's see: $\frac{1}{6} + \frac{1}{6} + \frac{1}{6}$ makes $\frac{3}{6}$, which is $\frac{1}{2}$. The result of this investigation is that $\frac{1}{6}$ is a *co-measurement* for $\frac{1}{2}$ and $\frac{1}{3}$, where a co-measurement of two fractions is a fraction that can be iterated to make both of the fractions. Using that co-measurement, $\frac{1}{2} + \frac{1}{3} = \frac{3}{6} + \frac{2}{6} = \frac{5}{6}$ (Figure 11.2).

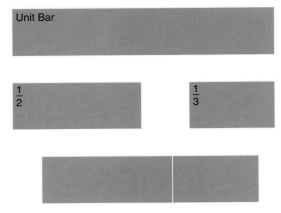

Figure 11.1 Adding $\frac{1}{2}$ of a unit and $\frac{1}{3}$ of the same unit

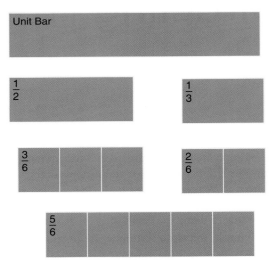

Figure 11.2 $\frac{1}{2} + \frac{1}{3} = \frac{3}{6} + \frac{2}{6} = \frac{5}{6}$

Although students at both Stages 2 and 3 can create equal size parts for adding and subtracting fractions (Steffe and Olive, 2010), as we have seen in Chapter 10, those at Stage 3 can create more general ways of thinking with fractions. In this chapter we focus on how students at Stages 2 and 3 can learn to add and subtract fractions.

Transforming Bars to Show Multiple Fractions

Before students engage in adding and subtracting fractions, it is helpful to work on creating multiple fractions in the same bar. For example, consider this problem:

> 'Transform Bars Problem A': Start with an energy bar that is partitioned into three equal parts. We'll call it a $\frac{3}{3}$-bar.
>
> 1) How can you transform that bar into a $\frac{21}{21}$-bar without erasing the thirds marks?
>
> 2) Colour your bar to show the thirds, and explain what you did.
>
> 3) How many twenty-firsts of the bar are the same amount as $\frac{1}{3}$ of the bar?
>
> 4) Can you see any other fraction in your $\frac{21}{21}$-bar? If so, colour your bar to show this fraction and explain how it is related to twenty-firsts.

The 'Transform Bars Problem A' represents a class of problems where the given number of parts in the bar is a factor of the targeted number of parts in the bar. When students partition each $\frac{1}{3}$ of the bar into 7 equal parts, the teacher has an opportunity to ask part (3) and introduce the idea of *commensurate fractions*. We view commensurate fractions as fractions that are the same size but created with a different number of parts (Steffe and Olive, 2010). Creating commensurate fractions requires that students have developed a Measurement Scheme for Unit Fractions and proper fractions, as discussed in Chapter 6. Creating commensurate fractions with many different fractions is a step in the development of equivalent fractions. We view equivalent fractions as an advanced concept in which a person thinks of a whole class of fractions as equal to $\frac{1}{3}$ (that is, fractions of the form $1n/3n$, where n is any whole number).

Problems like the 'Transform Bars Problem A' are accessible to students at both Stages 2 and 3, although part (4) of the problem will be more accessible to students at Stage 3 because these students can flexibly switch between viewing the candy bar as three 7-part segments and seven 3-part segments (Hackenberg, 2010; Steffe and Olive, 2010). That is, we expect it will be easier for Stage 3 students to propose that the 21 parts in the bar could also be organized into 7 equal parts, each containing 3 equal parts (see Figure 11.3). This organization can yield conversations about how $\frac{3}{21}$ is a commensurate fraction with $\frac{1}{7}$.

A more advanced transformation of bars is represented by this problem:

> 'Transform Bars Problem B': Start with an energy bar that is partitioned into two equal parts. We'll call it a $\frac{2}{2}$-bar.
>
> 1) How can you transform that bar into a $\frac{3}{3}$-bar without erasing the half mark?
>
> 2) Colour your bar to show the thirds, and explain what you did.

The 'Transform Bars Problem B' represents a class of problems where the given number of parts in the bar is not a factor of the targeted number of parts in the bar. Students at both Stages 2 and 3 can work on these problems, but students at Stage 2 are likely to find them more difficult,

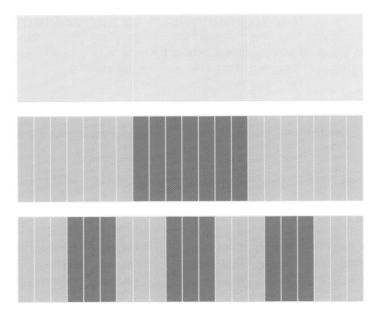

Figure 11.3   Reorganizing three-thirds each containing 7 equal parts into 7 equal parts each containing 3 equal parts

with some of the same patterns we discussed in Chapter 9. That is, solving the 'Transform Bars Problem B' is much like the problem of sharing two identical bars equally among three people, which is a considerable challenge for Stage 2 students.

The crux of the challenge is that to solve the 'Transform Bars Problem B' students have to think about the bar as two equal parts each containing an equal number of mini-parts, and then rearrange that total number of mini-parts into three equal parts each containing an equal number of mini-parts. Hackenberg (2010) has found that some Stage 2 students will learn to solve this problem, but they do not necessarily develop a general way of thinking in the way that Stage 3 students do.

Pulling Two Fractions Out of the Same Bar with a Co-Measurement

The transforming bars problems are related to problems of pulling out two fractions from the same bar using a co-measurement. Introducing the idea of a co-measurement can come out of a discussion of the transforming bars problems. In discussing these problems, the teacher can help students identify that $\frac{1}{21}$ is a co-measurement for $\frac{1}{21}$ and $\frac{1}{3}$ because it can be used (iterated) to make, or measure, both fractions: $\frac{1}{21} \times 1 = \frac{1}{21}$, and $\frac{1}{21} \times 7 = \frac{1}{3}$. Similarly, $\frac{1}{6}$ is a co-measurement for $\frac{1}{2}$ and $\frac{1}{3}$ because it can be iterated to make both fractions: $\frac{1}{6} \times 2 = \frac{1}{3}$ and $\frac{1}{6} \times 3 = \frac{1}{2}$. Again, these ways of thinking rely on having developed at least a Measurement Scheme for Unit Fractions and proper fractions, as discussed in Chapter 6.

Once students have developed the idea of a co-measurement as a fraction that can be used to measure two (or more) fractions, problems like the following can move them directly into adding and subtracting fractions:

'Pulling Out Two Different Sandwich Shares Problem': The rectangle represents a sub sandwich. Rita wants $\frac{1}{5}$ of the sandwich and Derrick wants $\frac{1}{3}$ of the sandwich.

1) What co-measurement can you use in order to pull out both Rita's and Derrick's shares from the same bar?

2) Draw and determine how much of the whole sandwich Rita and Derrick get together.

3) Write some mathematical notation to show your thinking on this problem. [This part of the problem can be further directed, depending on where the students are. For example, the teacher might say 'Write some mathematical notation starting with $\frac{1}{3} + \frac{1}{5}$ to show your thinking on this problem.']

Some students will make eighths in the bar in an effort to coordinate the idea of five parts and three parts. If they persist in this type of solution, we suggest moving back to the 'Transforming Bars Problems' to help students develop ideas about viewing multiple fractions in the same bar and about creating commensurate fractions. Students who create fifteenths will most likely be able to determine that together Rita and Derrick get 8/15. In writing notation in part (3), we suggest something like this: $\frac{1}{3} + \frac{1}{5} = \frac{5}{15} + \frac{3}{15} = \frac{8}{15}$. We do not suggest writing notation like $\frac{1}{3} \times \frac{5}{5} + \frac{1}{5} \times \frac{3}{3}$ until after focused work on commensurate fractions (discussed in the next section).

Work on problems like the 'Pulling Out Two Different Sandwich Shares Problem' can lead into a discussion of the idea that many co-measurements can be used to measure Rita's and Derrick's shares. For example, to solve the problem some students might use fifteenths and some might use thirtieths. Student use of different co-measurements can provide an opportunity to ask what other co-measurements can be used and what they have in common. As students list co-measurements, they may be able to articulate that the number of parts must be multiples of both 5 and 3 (or, both 5 and 3 must be factors of the number of parts used). Making this list provides a good opportunity to talk about why we might not use, for example, four-hundred fiftieths: it is easier to use fifteenths or thirtieths because the parts are bigger and so the number of parts needed is smaller. Students at both Stages 2 and 3 can articulate these ideas with supportive questioning, but students at Stage 3 are likely to make more general statements.

Here is a similar problem to the 'Pulling Out Two Different Sandwich Shares Problem' but with subtraction:

'Pulling Out Two Different Pizza Shares Problem': The rectangle represents a pizza. Steven wants $\frac{1}{7}$ of the pizza and Kyung wants $\frac{1}{4}$ of the pizza.

1)   What co-measurement can you use in order to pull out both Steven's and Kyung's shares from the same bar?

2)   Who gets more pizza, Steven or Kyung? How much more? Show this in your picture.

3)   Write some mathematical notation to show your thinking on this problem.

We note that some students interpret the 'pulling out two different shares' problems in this section to involve fraction multiplication. That is, some interpret the problem as Kyung wanting $\frac{1}{4}$ of Steven's piece (or Derrick wanting $\frac{1}{3}$ of Rita's piece). This interpretation can be a foundation for a discussion about referent units and how important it is to be explicit about that with fractions: Kyung wants $\frac{1}{4}$ of the pizza, $\frac{1}{4}$ of the rectangle shown, not $\frac{1}{4}$ of Steven's slice, which is $\frac{1}{7}$ of the rectangle shown. Extensions of these kinds of problems include working with more than just two fractions – see the instructional activities at the end of the chapter.

Commensurate Fractions More Generally

In order to develop general ways of adding and subtracting fractions, it is helpful for students to work on developing and **generalizing** their ideas about commensurate fractions.

In making commensurate fractions, one issue is how students generate many fraction names for the same size fraction using smaller and smaller parts. For example, students can partition parts to visualize that the following fractions are all the same size as $\frac{3}{4}$: $\frac{6}{8}$, $\frac{9}{12}$, $\frac{12}{15}$, etc. (Figure 11.4).

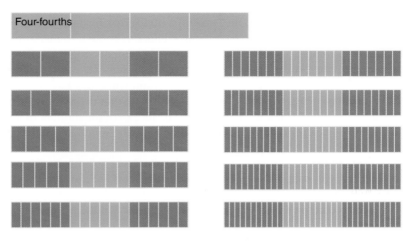

Figure 11.4   Ten fractions commensurate with $\frac{3}{4}$

In discussion teachers can help students to articulate relationships between each fraction and the original fraction, $\frac{3}{4}$. For example, in creating $\frac{6}{8}$ we have used parts that are one-half the size, and therefore we need twice as many parts to create the same-size fraction. Similarly, in creating $\frac{9}{12}$ we have used parts that are one-third the size, and therefore we need three times as many parts to create the same-size fraction. More generally, to make any fraction commensurate with $\frac{3}{4}$ we can make parts $1/k$ of the size, and so we need k times the number of parts to keep the same-size fraction.

This kind of discussion can lead into sensible notation, such as:

$$\frac{3}{4} \times \frac{1}{2} \times 2 = \frac{6}{8}$$

Here, multiplying by $\frac{1}{2}$ means that we are making parts one-half the size of the fourths, and multiplying by 2 means we need twice as many parts to create the same-size fraction. We prefer this kind of justification to the idea that we are multiplying $\frac{3}{4}$ by a form of 1 (i.e. $\frac{1}{2} \times 2 = 1$) because there is no quantitative reason to multiply $\frac{3}{4}$ by 1. Furthermore, the explanation that we are not changing $\frac{3}{4}$ because we are multiplying by a form of 1 does not indicate at all what is happening to the size of the parts and the number of parts used in the commensurate fractions.

A second issue in making commensurate fractions is the practice commonly called 'reducing fractions'. We know that often a lot of time and energy is spent in maths classes on this practice, although the name 'reducing fractions' is wildly misleading because it implies that the fraction is getting smaller in size. A better phrase is 'expressing fractions in lower or lowest terms'. To understand what is happening quantitatively when we express a fraction in lower terms, let's take the fraction $\frac{12}{30}$ (Figure 11.5).

Figure 11.5 $\frac{12}{30}$

In our view, to express $\frac{12}{30}$ in lower terms, we have to attend to sizes of parts and numbers of parts. We have to think about what size fractional parts we can use to make the same-size fraction, where we want to use a smaller number of parts than 30 in the whole. So, the question becomes: What number of parts is possible? Let's call the thirtieths 'mini-parts'. If we unite two mini-parts at a time to make a larger part, then we create fifteenths in the whole. Since fifteenths are twice as big as thirtieths, we only need one-half of the number of parts to create the same-size fraction. One-half of 12 is 6, so the fraction is $\frac{6}{15}$ (Figure 11.6).

Figure 11.6   Creating $\frac{6}{15}$ from $\frac{12}{30}$

But, could we use even bigger parts? If we unite three mini-parts at a time in the original $\frac{30}{30}$-bar, we create tenths in the whole. Since tenths are three times as big as thirtieths, we only need one-third of the number of parts to create the same-size fraction. One-third of 12 is 4, so the fraction is $\frac{4}{10}$ (Figure 11.7).

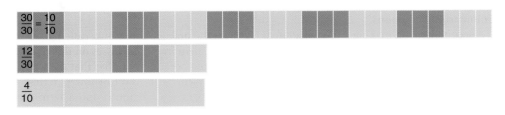

Figure 11.7   Creating $\frac{4}{10}$ from $\frac{12}{30}$

As teachers, we know we could still do more: We could unite six mini-parts at a time in the original $\frac{30}{30}$-bar to make fifths in the whole. Since fifths are six times as big as thirtieths, we only need one-sixth of the number of parts to create the same-size fraction. One-sixth of 12 is 2, so the fraction is $\frac{2}{5}$ (Figure 11.8).

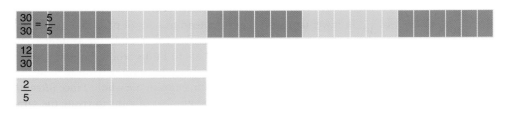

Figure 11.8   Creating $\frac{2}{5}$ from $\frac{12}{30}$

This activity can provoke some interesting discussion from students, who, after uniting three mini-parts at a time in the $\frac{30}{30}$-bar, may suggest that we unite four mini-parts at a time. Why is that not possible? Because we cannot fit larger parts consisting of four mini-parts into 30 mini-parts a whole number of times. In other words, 4 is not a factor of 30. Some students may say: What about five mini-parts? We can unite five mini-parts at a time to create six equal large parts in the whole. The only difficulty here is that we cannot do the same for the $\frac{12}{30}$-bar. That is, if we unite five mini-parts at a time within the 12 mini-parts that make up $\frac{12}{30}$, we come up with two equal large parts and two mini-parts. We do not come up with only equal large parts. In other words, 5 is not a factor of 12.

From this exploration two criteria emerge for expressing fractions in lower or lowest terms: (1) the number of parts united must be a factor of the total number of parts in the whole, and (2) the number of parts united must also be a factor of the number of parts in the fraction we are working with.

Furthermore, from this exploration we see that expressing fractions in lower or lowest terms means finding larger fractional parts with which to measure the given fraction. Since the parts are larger, we need multiplicatively fewer of them. So, rather than referring to changing the size of the fraction, the word 'reducing' refers to using a reduced number of parts to measure the same-size fraction. Nevertheless, we recommend not using the term 'reducing' for this process because of the connotations from that word that the fraction is getting smaller. Instead, we think it fruitful to emphasize that the fraction is exactly the same size, but we are finding larger, fewer parts with which to measure it.

Notation such as the following can trace the ideas above:

$$\frac{12}{30} \times 2 \times \frac{1}{2} = \frac{6}{15}$$

Here, multiplying by 2 refers to putting two parts at a time together to make parts twice as big. Then we need one-half of the parts that size to create the same-size fraction. This notation may be a challenge because students may not see multiplying $\frac{12}{30}$ by 2 as creating parts twice as big. If they do not, teachers can ask students to draw or imagine $\frac{1}{30} \times 2$. $\frac{1}{30} \times 2$ is $\frac{2}{30}$, which is the same as $\frac{1}{15}$. So, the multiplication by 2 does create parts twice as big.

Assessment Task Groups

List of Assessment Task Groups

A11.1: Transforming Bars

A11.2: Pulling Two Fractions from the Same Bar

A11.3: Adding Fractions

A11.4: Naming Fractions the Same Size

(Continued)

(Continued)

Task Group A11.1

Transforming Bars

Materials: Multiple copies of a piece of paper with copies of rectangles aligned on it, all partitioned into three equal parts (see Figure 11.9); coloured pencils or markers.

Figure 11.9   Multiple copies of the same rectangle, aligned

What to do and say: Point to the first bar and say, 'This is an energy bar partitioned into three equal parts. So we can call it a "three-thirds bar", written like this: $\frac{3}{3}$ -bar.' Then ask:

1)   Can you transform the bar into a $\frac{15}{15}$ -bar, that is, into a bar with 15 equal parts? Use one of the copies to show how.

2)   If the student partitions each of the thirds further, ask the student to colour the bar to show the thirds. If the student does so with confidence, proceed to (3) and (4).

3)   How many fifteenths of the bar are the same amount as $\frac{1}{3}$ of the bar? How do you know?

4)   Can you see any other fraction in your $\frac{15}{15}$ -bar? If so, use another copy of the bar, and colour it to show this fraction. Explain how this fraction is related to fifteenths.

Notes

- The free online applet *JavaBars* software is very useful for this problem (see http://math.coe.uga.edu/olive/welcome.html).
- If students struggle with this problem, you can use a number of parts that is smaller than fifteenths, i.e. sixths, ninths, or twelfths.
- If thirds and fifteenths seem to be easy for students, try starting with fourths or sixths and increase the targeted number of equal parts to an appropriate multiple. You can also ask students about fractions commensurate with a non-unit fraction. For example, a good follow-up to (3) is to ask them how many fifteenths are the same amount as $\frac{2}{3}$.
- This problem is like 'Transform Bars Problem A'. An extension of this problem is a problem like 'Transform Bars Problem B', in which the targeted number of equal parts is not a multiple of the given number of equal parts. Start with a $\frac{3}{3}$-bar, and ask students to create a $\frac{2}{2}$-bar. More challenging tasks include starting with a $\frac{2}{2}$-bar and creating a $\frac{3}{3}$-bar, or starting with a $\frac{3}{3}$-bar and creating a $\frac{4}{4}$-bar or $\frac{5}{5}$-bar. Always ask students to colour the bar to show the target fraction, and to say how they knew what to do.

Task Group A11.2

Pulling Two Fractions from the Same Bar

Materials: Multiple copies of a piece of paper with copies of rectangles aligned on it, all unmarked (so, like Figure 11.9, but unmarked); coloured pencils or markers.

What to do and say: Tell students that the rectangle represents a tray of clay. Two students, Tiana and Krista, are playing with the clay. Tiana wants $\frac{1}{3}$ of the tray, and Krista wants $\frac{1}{4}$ of the tray. How can students partition the tray into one fraction family to give Tiana and Krista each what they need? In other words, students can partition the tray into thirds and then stop partitioning. Can they give both girls their clay amounts now? (No.) Or the students can partition the tray into fourths and then stop partitioning. Can they give both girls their clay amounts now? (No.) What partitioning will work? How does it work?

Notes

- The reason to use the term 'one fraction family' and give the examples above is that, since this is an assessment, students will likely not have learned about the idea of a co-measurement.
- Some students try sevenths, for natural reasons – they are trying to coordinate 3 and 4 and they add. However, they will usually see that this does not allow them to take out the clay amount for either girl.
- If thirds and fourths seem too difficult, you can start with thirds and halves.
- If students are successful in pulling out both parts, then you can ask the following:

  o How much clay do the girls have together? How do you know? Can you write an addition sentence to represent what you did?

(Continued)

(Continued)

    ○   Who has more clay? How do you know? How much more? Can you write a subtraction sentence to represent what you did?

- You can extend this assessment by using proper fractions. If you use two proper fractions that together are larger than the whole tray, you can tell students that the tray holds 'magic' clay. The clay is magic because if you take a part out, it fills right back in.

Task Group A11.3

Adding Fractions

Materials: Multiple copies of a piece of paper with multiple copies of $\frac{3}{3}$-bars on it (Figure 11.9); writing utensils.

What to do and say: Ask students, 'Can you use these bars to show how to add one-third and one-twelfth?' Ask them to explain their solution, if they make one.

Notes

- The purpose of this task is to see whether or not students will use the idea of finding a co-measurement to solve the problem.
- The task can be changed in various ways. You can use it to assess whether students can add fractions that are already the same size parts, although then you may want to use a larger number of parts in the bars, such as $\frac{5}{5}$-bars. Ask: 'Can you use these bars to show how to add two-fifths and one-fifth?'
- To increase the challenge, pose problems that surpass the whole, such as $\frac{2}{3} + \frac{5}{12}$.
- To increase the challenge, pose problems that involve smaller parts, such as $\frac{1}{6} + \frac{5}{24}$.
- To increase the challenge, pose problems where the co-measurement of the two fractions is not one of the fractions, such as $\frac{1}{3} + \frac{3}{4}$.

Task Group A11.4

Naming Fractions the Same Size

Materials: Multiple copies of a piece of paper with copies of rectangles aligned on it, all unmarked (so, like Figure 11.9, but unmarked); coloured pencils or markers.

What to do and say: Ask students to draw the fraction $\frac{2}{5}$. Then ask them to name all the fractions they know that are the same size as $\frac{2}{5}$. They can list them verbally or by writing fraction notation. Ask them how they know a particular fraction is the same size as $\frac{2}{5}$; can they draw a picture to show how they know?

Notes

- The purpose of this activity is to see whether students will name other fractions and whether they have imagery about sizes to justify these other fraction names. Some students who have experienced some fraction instruction may name $\frac{4}{10}$ as the same as $\frac{2}{5}$ because they multiplied the numerator and denominator each by 2, but they may have no imagery developed to understand why these two fractions are the same amount.
- To decrease the complexity of the task, you can use unit fractions; to increase it, use proper fractions with more parts (e.g. $\frac{5}{7}$).

Instructional Activities

List of Instructional Activities

IA11.1: Transforming Bars I

IA11.2: Transforming Bars II

IA11.3: Add and/or Subtract

IA11.4: Making Same-Size Fractions with Smaller Parts

IA11.5: Making Same-Size Fractions with Larger Parts

Activity IA11.1

Transforming Bars I

Intended learning: Students will learn how to partition a given partitioned bar in order to show a different fraction in the bar, where the number of parts in the target fraction is a multiple of the number of parts in the given bar.

Instructional mode: Students working in pairs with instruction from the teacher, followed by whole-class discussion.

Materials: Multiple copies of a piece of paper with $\frac{3}{3}$-bars on it (see Figure 11.9), $\frac{4}{4}$-bars, $\frac{5}{5}$-bars, etc. Depending on class size you might want each pair of students to get a differently partitioned set of bars, or you can have some pairs use the same. Additional materials include coloured pencils or markers, and fractions written on small index cards or pieces of paper. The fractions should be out of a number of parts that is a multiple of the number of parts in the bars. For example, if you are using $\frac{5}{5}$-bars, you might have $\frac{1}{10}, \frac{1}{15}, \frac{1}{20}, \frac{1}{25}$, etc.

(Continued)

(Continued)

Description: Give each pair of students some paper with rectangles that have the same number of parts. Also slip to each student in the pair a different fraction to make. For example, if a pair of students had the $\frac{5}{5}$-bars, then you could give one student $\frac{1}{10}$ and the other student $\frac{1}{15}$. They should not know each other's fractions initially. Ask each student to transform a bar to show their fraction but without showing it to their partner. Then partners tell each other what their target fraction was and predict what their partner did to make it. Following the predictions, partners can show each other their drawings to check. Finally, they should each colour their bars to show the original fraction and state how many of the target fractions are needed to make, or measure, one unit fraction of the original bar. For example, how many $\frac{1}{10}$ s are the same size as $\frac{1}{5}$? Then they each work with another fraction. Following several rounds, hold a whole-class discussion about how students made their target fractions and what they noticed about how many target fractions made up a unit fraction of the original bar. Here you can also introduce the term 'co-measurement'.

Responses, Variations and Extensions

- To increase the challenge, you can ask students how to show both of their target fractions in the same bar. For example, how can you show $\frac{1}{10}$ and $\frac{1}{15}$ in the same bar? Students should colour their bar to show both of the original target fractions. If they work successfully on this challenge, they are likely ready for IA11.2.
- This activity could be done with *JavaBars*.
- This activity could be played as a game, where students earn points for correct predictions and correct solutions.

Activity IA11.2

Transforming Bars II

Intended learning: Students will learn how to partition a given partitioned bar in order to show a different fraction in the bar, where the number of parts in the target fraction is not a multiple of the number of parts in the given bar.

Instructional mode: Students working in pairs with instruction from the teacher, followed by whole-class discussion.

Materials: Multiple copies of a piece of paper with $\frac{2}{2}$-bars, $\frac{3}{3}$-bars and $\frac{4}{4}$-bars; coloured pencils or markers; fractions written on small index cards or pieces of paper where the number of parts in the target fraction is not a multiple of the number of parts in the given bar.

Description: This activity is a good deal harder than IA11.1, so we recommend that students work together on one fraction to come up with ideas. That is, give each pair of students some paper with rectangles that have the same number of parts and a fraction to make, where they are not to erase or ignore the marks on the rectangles. For example, one pair might have $\frac{2}{2}$-bars and the fraction $\frac{1}{3}$. So, their task would be to transform the $\frac{2}{2}$-bar into a $\frac{3}{3}$-bar without erasing the half mark. Students can present and explain their solutions to the class. Discussion can focus on how they transformed their bars and what they noticed about the co-measurement that they found for the two different fractions. You may also want to contrast IA11.1 with IA11.2.

Responses, Variations and Extensions

- Students may not come up with the same ideas here. For example, to change a $\frac{2}{2}$-bar into a $\frac{3}{3}$-bar, one student might mark each half into six equal parts, and another might use three equal parts in each half. Differences like this provide a good opportunity to talk about how many ways there are to solve the problem. For example, how many different co-measurements are there for halves and thirds and what do they have in common? As we discussed in this chapter, it may also be useful to ask why they might not have used sixtieths to solve the problem (it would work, but the parts would be so little that it would be hard to make).
- This activity could be done with *JavaBars*.
- This activity could be played as a game, where students earn points for correct predictions and correct solutions. However, because it is more challenging it may be best to hold off on changing it to a game for some time. If you do play it as a game, see IA11.1 for ways that partners can each work on their own fraction.

Activity IA11.3

Add and/or Subtract

Intended learning: Students will learn how to pull two fractions out of the same bar in order to add or subtract them.

Instructional mode: Students working in pairs or small groups with instruction from the teacher.

Materials: Multiple copies of a piece of paper with copies of rectangles aligned on it, all unmarked (so, like Figure 11.9, but unmarked); coloured pencils or markers.

Description: Pose problems where you ask students to add two fractions of a quantity, like the 'Pulling Out Two Different Sandwich Shares Problem', or subtract two fractions of a quantity, like the 'Pulling Out Two Different Pizza Shares Problem'. Students can present their solutions, and you can engage students in writing notation for their ideas as we have discussed in this chapter. You can also pose problems that involve a unit of measure, such as: 'Mrs Garcia was getting new sand for the school play-ground. She had $\frac{2}{3}$ of a tonne of sand delivered, but it was not enough, so she bought $\frac{1}{4}$ of a tonne more. How much sand does she have to use for the playground?' Other useful measures for these problems include tonnes of soil, gallons of paint, pounds or kilogrammes of flour or other ingredients, and ounces of spices. We recommend starting with unit fractions as in the two problems above from this chapter, and then moving to non-unit fractions.

(Continued)

(Continued)

Responses, Variations and Extensions

- After students have worked on adding two fractions and subtracting two fractions, pose problems like this one: 'Katy has $\frac{4}{5}$ of a gallon of petrol in her lawnmower. She buys $\frac{1}{4}$ of a gallon more from her neighbour. Then she uses $\frac{3}{8}$ of a gallon to mow the lawn. How much of a gallon does she have now?' In this problem, students have to add and subtract, and they have to make a co-measurement for three different fractions.
- This activity could be done with *JavaBars*, which is especially useful for working with a large number of parts within a bar, as may be needed when combining multiple fractions additively.

Activity IA11.4

Making Same-Size Fractions with Smaller Parts

Intended learning: Students will learn how to create commensurate fractions with smaller parts, and therefore with more parts.

Instructional mode: Students working in pairs or small groups with instruction from the teacher.

Materials: Paper with a $\frac{5}{5}$-bar on it and multiple $\frac{2}{5}$-bars drawn beneath, with the mark between the two $\frac{1}{5}$s in all $\frac{2}{5}$-bars well visible; writing utensils.

Description: Tell students that the $\frac{5}{5}$-bar represents a litre of apple cider, and below is $\frac{2}{5}$ of a litre of apple cider. Ask students to draw 8 or 10 fractions of a litre of cider that are equal to $\frac{2}{5}$ of a litre. If students work with a partner or small group then they can each draw a few, but they should talk with their group-mates first about what they plan to draw. As students work, most likely they will produce $\frac{4}{10}$ of a litre by partitioning each fifth into two equal parts. Engage students in a whole-classroom discussion about how the parts of $\frac{4}{10}$ compare to the size of the parts of $\frac{2}{5}$, and about how many parts are needed to make $\frac{4}{10}$ compared with the number of parts needed to make $\frac{2}{5}$. Engage students in a similar discussion about $\frac{6}{15}$, as well as another fraction they have made (maybe $\frac{8}{20}$). Then ask them what is happening to the sizes of parts and numbers of parts as they make these fractions. In the discussion you can write notation to show the process of transforming $\frac{2}{5}$ to $\frac{4}{10}$, similar to what is written in this chapter: $\frac{2}{5} \times \frac{1}{2}$ makes the parts one-half the size, but then you need twice as many parts to maintain the same-size fraction, so $\frac{2}{5} \times \frac{1}{2} \times 2 = \frac{4}{10}$.

Responses, Variations and Extensions

- Of course, other fractions can be used as a basis for this activity.
- This activity could be done with *JavaBars*, which is particularly advantageous for this activity!

Activity IA11.5

Making Same-Size Fractions with Larger Parts

Intended learning: Students will learn how to create commensurate fractions with larger parts, and therefore with fewer parts.

Instructional mode: Students working in pairs or small groups with instruction from the teacher.

Materials: Paper with multiple pictures of an $\frac{18}{18}$-bar with a $\frac{12}{18}$-bar beneath; writing utensils.

Description: Tell students that the $\frac{12}{18}$-bar represents part of whole pumpkin bread that a bakery is selling. The store manager has many other loaves of pumpkin bread, and she wants to cut the next loaf into fewer pieces but still be able to put $\frac{12}{18}$ of the loaf in the display window. In other words, she wants to make pieces that are bigger than eighteenths. Ask students to show at least two different ways that she could do this and to explain their solutions. When students present their solutions to the class, focus them on what they are doing with the parts to create the larger pieces and to use fewer of them; don't worry about mathematical notation initially. Ask students to articulate what is happening to the size of the parts and the number of parts in their solutions. Then, in the discussion you can help them to write notation to reflect this process. For example, $\frac{12}{18} \times 3$ represents making parts three times the size (sixths), and then just one-third of those parts are needed, so $\frac{12}{18} \times 3 \times \frac{1}{3} = \frac{4}{6}$.

Responses, Variations and Extensions

- If it does not come up in the discussion, you could suggest that a student named Emily wants to put 4 parts of the $\frac{18}{18}$-bar together at a time in the $\frac{12}{18}$-bar. Will this work? (No, for reasons we discuss in this chapter.) Similarly, you could ask students whether it is possible to unite nine parts into a larger part.
- Of course, other fractions can be used as a basis for this activity. However, it is helpful to start with numbers of parts that have several factors in common. For example, 12 and 18 have 1, 2, 3 and 6 in common.
- This activity could be done with *JavaBars*; it is particularly advantageous for this activity!
- In general, instructional activities in this chapter address standards related to equivalent fractions, such as Common Core State Standards 3.NF.3, 4.NF.1, and 5.NF.1, as well as standards related to fraction addition, such as Common Core State Standards 4.NF.3, 5.NF.1 and 5.NF.2.

12

Teaching Students at Stages 2 and 3: Dividing Fractions

Domain Overview

Whole number division can have a partitive meaning, in which a known quantity is to be shared into a known number of equal shares and the size of those shares is not known. It can also have a quotitive (measurement) meaning, in which a known quantity is to be shared into shares of a known size, and the number of equal shares is unknown. This latter meaning with fractions is the focus of this chapter; we address the other meaning of division with fractions in Chapter 13, which focuses on fractions and algebraic reasoning.

Just as with fraction multiplication (Chapter 10) and fraction addition and subtraction (Chapter 11), students at both Stages 2 and 3 can work on developing a measurement meaning for dividing fractions. However, students at Stage 3 can create more general ways of thinking.

Fraction Division Problems with a Measurement Meaning

Here is an example of a fraction division problem with a measurement meaning:

'Flowerbed Problem': A landscape architect bought 5 tonnes of soil, knowing that each flowerbed in the garden she was designing required $\frac{3}{5}$ of a tonne of soil. How many flowerbeds can she fill with soil?

1)   Solve this problem with a drawing. Include fractional amounts of flowerbeds if needed.

2)   To answer this problem, Beth got $8\frac{1}{5}$ flowerbeds, while Tyrone got $8\frac{1}{3}$ flowerbeds. They have a debate about who is correct. Who is correct, and why?

3)   This problem can be viewed as a division problem. Why?

4)   Write a division sentence for your work on this problem.

Students at both Stages 2 and 3 often work on part (1) of this problem by drawing a picture of 5 tonnes of soil. Then they partition each tonne into five equal parts, count off three $\frac{1}{5}$ tonnes at a time, and determine that they have made eight $\frac{3}{5}$-tonne shares, with one $\frac{1}{5}$-tonne part left over (Figure 12.1).

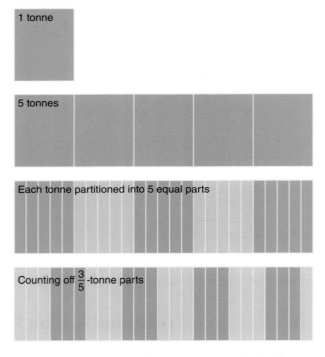

Figure 12.1   A drawing and counting solution for the 'Flowerbed Problem'

Some students at Stage 3 may reason like this: There are at least five $\frac{3}{5}$-tonnes in 5 tonnes because there is one $\frac{3}{5}$-tonne inside each tonne. Then there are five $\frac{2}{5}$-tonnes left over, which is $\frac{10}{5}$ of a tonne. That can make three more $\frac{3}{5}$-tonne shares, and so there are a total of 5 + 3, or 8, $\frac{3}{5}$-tonne shares in 5 tonnes, with one $\frac{1}{5}$-tonne part left over (Figure 12.2). This reasoning involves strategically organizing the fifths within the tonnes, which can be quite challenging for students at Stage 2.

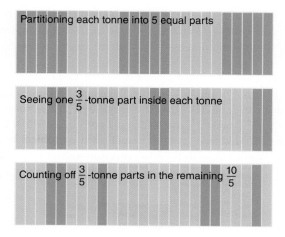

Figure 12.2 Another solution pathway: Seeing one $\frac{3}{5}$-tonne part in each of the 5 tonnes

With either solution, one $\frac{1}{5}$-tonne part is left over. So, the question becomes how to view that part. Many students at both stages will think the answer is $8\frac{1}{5}$ flowerbeds. Some will specify that the answer is 8 flowerbeds with $\frac{1}{5}$ tonne of soil left over. Most likely a small number of students will think the answer is $8\frac{1}{3}$ flowerbeds because $\frac{1}{5}$ tonne of soil can make $\frac{1}{3}$ of a flowerbed, since $\frac{1}{5}$ tonne is $\frac{1}{3}$ of the $\frac{3}{5}$-tonne needed for a whole flowerbed. Part (2) of the problem can induce students to understand these different responses to the problem and to realize that $8\frac{1}{5}$ flowerbeds is not correct.

Some Stage 2 students may struggle with this aspect of the problem because there are several different units involved: the unit tonne, $\frac{1}{5}$ of a tonne, $\frac{3}{5}$ of a tonne. A key idea to come out of the discussion is that when we are measuring off amounts of soil for flowerbeds, we are actually naming the result in relation to $\frac{3}{5}$ of a tonne as the unit of measurement for the 5 tonnes. So, $\frac{3}{5}$ of a tonne as a unit fits into 5 tonnes 8 times and $\frac{1}{3}$ of a time, or $8\frac{1}{3}$ times.

This idea is a foundation for responding to part (3), which may require some review of different meanings for whole-number division. Students at Stages 2 and 3 can learn to articulate that this problem is a division problem because it can be seen as repeatedly measuring off $\frac{3}{5}$ of a tonne from 5 tonnes to see how many shares of that size can be made. In other words, the problem can be seen as asking how many times $\frac{3}{5}$ of a tonne fits into 5 tonnes, and it can be notated $5 \div \frac{3}{5} = 8\frac{1}{3}$ (part 4).

A Multiplicative Solution to Fraction Division Problems with a Measurement Meaning

Drawing and counting solutions for measurement division problems are accessible to students and helpful for them to develop. However, they are cumbersome with larger numbers, and they are not as efficient as multiplicative solutions. So, how could a person solve a problem like the 'Flowerbed Problem' without drawing and counting up each share of known size? We will sketch a pathway for a multiplicative solution here and then discuss ideas that are needed to build up to that solution.

In the 'Flowerbed Problem', the architect needs $\frac{3}{5}$ of a tonne of soil to make 1 flowerbed. However, since she has 5 tons of soil, it would really be more useful to know how many flowerbeds can be made with 1 tonne of soil. As shown in Figure 12.3, since $\frac{3}{5}$ of a tonne makes 1 flowerbed, 1 whole ton will make 1 flowerbed and 2 parts more. Those 2 parts must be thirds of a flowerbed, because a flowerbed consists of 3 equal parts, each consisting of $\frac{1}{5}$ tonne of soil. So, 1 tonne of soil will make 1 flowerbed and $\frac{2}{3}$ of a flowerbed, or $1\frac{2}{3}$ flowerbeds.

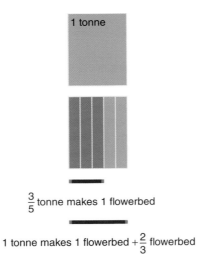

$\frac{3}{5}$ tonne makes 1 flowerbed

1 tonne makes 1 flowerbed $+\frac{2}{3}$ flowerbed

Figure 12.3   The relationships between tonnes of soil and flowerbeds

Determining the number of flowerbeds that 5 tonnes of soil make involves multiplying the number of flowerbeds for 1 tonne of soil by 5. So, $5 \times 1\frac{2}{3}$ should give the total number of flowerbeds. One way to multiply $5 \times 1\frac{2}{3}$ is to multiply 5×1 and $5 \times \frac{2}{3}$. That yields 5 flowerbeds and $\frac{10}{3}$ flowerbeds, which is $8\frac{1}{3}$ flowerbeds.

Here we have changed the division problem, $5 \div \frac{3}{5}$ into a product, $5 \times 1\frac{2}{3}$. So, $5 \div \frac{3}{5} = 5 \times 1\frac{2}{3}$. Students who have developed fractions as numbers (see Chapter 8) will see that $1\frac{2}{3} = \frac{5}{3}$, and so $5 \div \frac{3}{5} = 5 \times \frac{5}{3}$. Teachers may recognize that this multiplicative solution is one example of the standard computational algorithm for fraction division with a measurement meaning. That is, to divide a quantity by $\frac{3}{5}$ we have multiplied the quantity by its reciprocal, $\frac{5}{3}$.

Notice that this solution requires flexibly shifting among many different units, including 1 tonne and 1 flowerbed (which is given to be equivalent to $\frac{3}{5}$ of a tonne of soil), as well as the potential use of improper fractions in the process of multiplying 5 and $1\frac{2}{3}$. In addition, conceiving of fractions as numbers is necessary to view the multiplicative solution as involving multiplication by a reciprocal. So, in our view, this kind of solution is in the purview of students at Stage 3, even though students at Stage 2 can work on the ideas to some extent.

A Quantitative Meaning for a Reciprocal

One critical idea in the move to multiplicative solutions of fraction measurement division problems is a quantitative meaning for a reciprocal. In our view, a quantitative meaning emphasizes relationships among quantities as opposed to solely the idea that a fraction and its reciprocal multiply to 1. To help develop a quantitative meaning for a reciprocal, consider the following problems:

'One-fourth of a Metre Problem': The following rectangle (or segment) has length $\frac{1}{4}$ of a metre.

1)   Use it to draw the whole metre.

2)   How many times does $\frac{1}{4}$ of a metre fit into 1 metre?

3)   What can you multiply $\frac{1}{4}$ of a metre by to make 1 metre?

'Two-thirds of a Mile Problem': The following rectangle (or segment) has length $\frac{2}{3}$ of a mile.

1)   Use it to draw the whole mile.

2)   How many times does $\frac{2}{3}$ of a mile fit into 1 mile?

3)   What can you multiply $\frac{2}{3}$ of a mile by to make 1 mile?

Students at both Stages 2 and 3 can work on such problems. However, often students at Stage 2 will develop only $1\frac{1}{2}$, not $\frac{3}{2}$, as a solution to part (3) of the 'Two-thirds of a Mile Problem', and they will find many other fractions quite difficult to work with. It is okay for students to think of the answer to parts (2) and (3) as a mixed number, but doing so limits the possibilities for students to become aware of patterns in their thinking. That is, after doing many problems like these, it can be useful to pose the following:

'Patterns in Our Reasoning Problem': After solving these problems, take a look at your solutions. Do you see any patterns in your responses to part (3)? How do your responses to (3) relate to the original quantity in the problems?

In discussing patterns that students notice in their reasoning, the teacher might introduce the term 'reciprocal' and help students to articulate its quantitative meaning as the number of times a given quantity fits into one unit of the quantity.

Problems to Elicit Multiplicative Solutions of Fraction Measurement Division Problems

After work on a quantitative meaning for a reciprocal, structuring problems as follows can help students work toward multiplicative solutions.

The following problem should look familiar because you have already solved it using a picture. Now we're going to develop a different solution:

'Flowerbed Problem, Take 2': A landscape architect bought 5 tonnes of soil, knowing that each flowerbed in the garden she was designing required $\frac{3}{5}$ of a tonne of soil. How many flowerbeds can she fill with soil?

1) Determine how many flowerbeds the architect can make with 1 tonne of soil. Show your reasoning clearly.

2) Use part (1) to determine how many flowerbeds she can make with 5 tonnes of soil.

3) Write a multiplication sentence for this situation, and explain what each number in your sentence means in terms of tonnes and flowerbeds.

4) In your first solution of this problem, you wrote a division statement: $5 \div \frac{3}{5} =$ _____ flowerbeds. Fill in this statement. How is your division statement in (4) related to your multiplication statement in (3)?

Doing many problems like these is important in developing patterns in reasoning. We recommend working with a known whole number quantity and a known fraction share, as in the 'Flowerbed Problem', for some time before moving to problems in which both numbers are fractions (e.g., the architect has $5\frac{1}{4}$ tonnes of soil and $\frac{2}{3}$ of a tonne makes a flowerbed).

Students at Stage 3 are well poised to make generalizations about these ideas. One way to view the generalization that students at Stage 3 can make is as follows: the known fraction share in a fraction measurement division problem gives, in effect, a unit ratio – the amount of material for 1 unit of another entity ($\frac{3}{5}$ of a tonne of soil makes 1 flowerbed). Finding the reciprocal quantitatively finds the other unit ratio: the amount of the entity for 1 unit of the material ($\frac{5}{3}$ flowerbeds are made with 1 tonne of soil). Then, to solve the problem requires scaling that other unit ratio by the known initial quantity (scaling 5/3 flowerbeds per 1 tonne by 5 because we have 5 tonnes of soil, not 1 tonne of soil).

Scaling some fractions by whole numbers is in the purview of students at Stage 2. For example, they might be able to solve the 'Flowerbed Problem, Take 2' by thinking about taking $1\frac{2}{3}$ flowerbeds five times. However, if the initial quantity was $7\frac{3}{4}$ tonnes, or $13\frac{5}{8}$ tonnes, or $\frac{19}{3}$ tonnes, the problem would be quite hard for students at Stage 2 to solve with meaning. In contrast, given appropriate instructional environments we expect students at Stage 3 to be able to develop these ideas with 'any' fraction (within reason).

Assessment Task Groups

List of Assessment Task Groups

A12.1:   Drawing Pictures of Whole Number Division

A12.2:   Meanings of Division

A12.3:   Writing Division Stories

A12.4:   Fitting into One

Task Group A12.1

Drawing Pictures of Whole Number Division

Materials: Paper and writing utensil.

What to do and say: Ask students to draw a picture to show what $24 \div 4$ means. Look for whether they draw 24 partitioned into 4 equal groups or 24 measured off into groups of size 4, or both. Ask them to tell you about their picture and what division means.

Notes

- Some students may have no idea about drawing a picture, or they may say only that division is the 'opposite' of multiplication.
- Students who draw 4 equal groups are demonstrating a partitive meaning for division, which is discussed in Chapter 13; students who draw groups of size 4 are demonstrating a measurement meaning of division.
- When prompted, students who demonstrate a partitive meaning might be able to draw another picture that demonstrates a measurement meaning. The teacher might ask whether they can think about it another way and draw a different picture. The teacher could even show them a picture representing the measurement meaning and ask them about it: 'Does this also show 24 divided by 4?'

Task Group A12.2

Meanings of Division

Materials: Paper with the following problems printed on it:

(1)   Sara the gardener has 48 square tiles. She wants to put them into 4 rows, with the same number of tiles in each row. How many tiles will be in each row?

(Continued)

(Continued)

(2) Sara the gardener has 48 square tiles. She wants to put them into rows with 4 tiles in each row. How many rows can she make?

(3) Sara the gardener is making rows of square tiles with 4 tiles in each row. She wants to make 48 rows. How many tiles does she need?

(4) Sara the gardener has 48 square tiles. She makes 3 rows with 4 tiles in each row. How many tiles does she have left after doing this?

What to do and say: Ask students to read each of these problems with you (or you can read the problems to younger students). Ask students which problems involve division and why. If they identify both (1) and (2) as involving division, you can also ask them what is the same and what is different about the two problems.

Notes

- These tasks are intended for the teacher to get a better understanding of students' meanings of division. The teacher can learn how robust those meanings are and also learn about some of the nuances in students' thinking. In addition, the teacher might learn how students' meanings of division relate to their meanings of multiplication and subtraction.

- These tasks also provide an opportunity for students to begin reflecting on their meanings and understandings of division.

Task Group A12.3

Writing Division Stories

Materials: Paper and writing utensils.

What to do and say: Ask students to think about the problem '8 divided by ½'. Ask them to develop a story problem that would be solved by '8 divided by ½'. After they have a story, they should read it and say why it is a division problem.

Notes

- Students may find this task quite challenging. If they write a story that results in 8 divided by 2, ask them about the answer to their problem. Then ask them if the answer to 8 divided by $\frac{1}{2}$ should be the same as the answer to 8 divided by 2. Be aware that some students will say yes!

- If students succeed at creating a story, most likely it will be a measurement division story, which involves determining the number of times $\frac{1}{2}$ fits into 8. As an extension, ask them if they could create a story for '3 $\frac{1}{2}$ divided by $\frac{1}{4}$'.

Task Group A12.4

Fitting into One

Materials: Multiple copies of a piece of paper with 2–3 different-sized bars on it, with space between them, e.g. see Figure 12.4.

Figure 12.4   Set-up for 'Fitting into One' task

What to do and say: Tell students that this first rectangle is $\frac{1}{3}$ of a fruit bar. Can they draw the fruit bar? How many times does $\frac{1}{3}$ of a fruit bar fit into 1 fruit bar? What can they multiply $\frac{1}{3}$ of a fruit bar by to make 1 fruit bar?

Notes

- If students solve the problem above, you can repeat the problem with a different unit fraction.
- To extend the problem, tell students that one of the other rectangles represents $\frac{2}{5}$ of a fruit bar and repeat the same questions. Starting with a fraction that is two fractional units ($\frac{2}{3}, \frac{2}{5}, \frac{2}{7}$, etc.) is helpful here because students may be able to see that some whole number of those parts and one-half of a part fit into 1 unit (e.g. $2\frac{1}{2}$ $\frac{2}{5}$-bars fit into 1 bar). To solve this problem, they need to have at least developed a splitting operation (see Chapter 7).
- If students find the questions with two fractional units easy, then you could repeat the task with $\frac{3}{4}$ of a fruit bar. They may find this question quite challenging because they have to see the remaining fourth in the whole fruit bar as one-third of the $\frac{3}{4}$-bar.

Instructional Activities

List of Instructional Activities

IA12.1:   Meanings of Division

IA12.2:   Writing Division Stories

IA12.3:   Solving Measurement Division Problems I

IA12.4:   Reciprocals

IA12.5:   Solving Measurement Divisions Problems II

Activity IA12.1

Meanings of Division

Intended learning: Students will learn to distinguish between two meanings for division.

Instructional mode: Students working in small groups with initial instruction from the teacher.

Materials: Long ribbons to represent bubble gum tape (two per group), ruler, scissors, paper.

Description: Give each group of students a length of ribbon that represents some number of inches or centimetres of bubble gum tape. For example, one group has 39 inches. A single serving is 3 inches. Ask them to find how many people can get a serving from their tape, to tell how they know, and to make the servings and draw a picture of the process of making the servings. Then ask them to write a division statement for what they did and explain how they see this statement in their picture. Then give them another 39 inches. But this time they are to share that 'gum' equally with 3 people. Ask them how much each person gets, to tell how they know, and to make the servings and draw a picture of the process of making the servings. Then ask them to write a division statement for what they did and explain how they see this statement in their picture. Then ask them to describe the differences between these two situations and the pictures and sentences they wrote. In a whole-class discussion, groups can present their work to the class. Identify with them that the first meaning of division is called a measurement meaning because they measured off a known size (3 inches) over and over from the tape. The second meaning is called a sharing meaning because they made a known number of equal shares (3) from the tape.

Responses, Variations and Extensions

- You can increase the difficulty level by increasing the numbers or by moving into fractions — extending the range of numbers and complexifying the arithmetic.
- For example, tell students they have 6 feet of bubble gum tape and a single serving is ½ foot. How many people can get a single serving? Ask them to draw a picture to develop their solutions.

Then write a division sentence and tell what it means in terms of the picture. Ask: What meaning of division are you using here?

- Or, tell students that they have 5 feet of bubble gum tape and four people want to share it equally. How much does each person get? Ask students to draw a picture to develop their solutions. Then to write a division sentence and say what it means in terms of the picture. Ask: What meaning of division are you using here?

Activity IA12.2

Writing Division Stories

Intended learning: Students will learn how to write stories that reflect different meanings of division.

Instructional mode: Students working in pairs with initial instruction from the teacher.

Materials: Paper, pencils, markers.

Description: Give each student in the pair a division statement that they won't show to their partner. For example, one student gets $8 \div \frac{1}{3}$ and one student gets $5 \div \frac{1}{4}$. Each student is to write a division story problem that would result in that division computation. Then they exchange stories. Each student reads their partner's story, uses a picture to try to solve the partner's problem, and writes the division statement that they think solves the problem. Then they check in with their partner to see if this fits with the original division statement.

Responses, Variations and Extensions

- This activity can be made into a game with students earning points for correctly writing stories to represent division statements and for correctly solving stories.
- This activity can be used with whole numbers.
- This activity can be expanded so that students initially get multiplication or division statements, or addition, subtraction, multiplication, or division statements.

Activity IA12.3

Solving Measurement Division Problems I

Intended learning: Students will learn how to use pictures to solve measurement division problems involving fractions.

Instructional mode: Students working in pairs or individually with initial instruction from the teacher.

Materials: Paper, pencils, markers, worksheet of problems like the 'Flowerbed Problem'.

(Continued)

(Continued)

Description: In this activity, students work on developing solutions via drawings for fraction measurement division problems. Students could work in pairs or small groups and present their solutions to the class. Here are three sample problems that could be used. In each case, it is helpful to use parts (1), (3) and (4) of the 'Flowerbed Problem', adapted for the particular problem. You can also use a version of part (2) for some or all of the problems in order to help students think about what to do with leftover parts. Since part (2) varies by problem, a sample part (2) has been included for each problem below.

I)   A grocer gets 7 kg of rice delivered. She has to package the rice into bags that contain $\frac{3}{4}$ kg. How many bags can she make, including fractions of a bag? Sample part (2): To answer this problem, Sahara got $9\frac{1}{4}$ bags, while Ben got $9\frac{1}{3}$ bags. Who is correct and why?

II) At a camp, a jug of lemonade contains $4\frac{1}{2}$ litres. The camp counsellor will give out $\frac{1}{3}$-litre servings of the lemonade to campers. How many servings can be given out, including fractions of a serving? Sample part (2): To answer this problem, Aviva got $13\frac{1}{2}$ servings, and Chris got $12\frac{1}{2}$ servings. They have a debate about who is correct. Who is correct and why?

III) A baker uses $\frac{2}{3}$ of a cup of sugar for each batch of cookies he makes. He has $5\frac{1}{6}$ cups of sugar. How many batches of cookies can he make, including fractions of a batch? Sample part (2): To answer this problem, Eli got $7\frac{1}{2}$ batches, while Dionne got $7\frac{3}{4}$ batches. They have a debate about who is correct. Who is correct and why?

Responses, Variations and Extensions

- Ideally, students can work on these problems in *JavaBars* (math.coe.uga.edu/olive/welcome.html).
- If students have developed fractions as numbers (Chapter 8), improper fractions can be used for the total amount as well as the known share size.
- These activities support standards that promote the use of visual models to solve real-world problems involving fraction division – standards like Common Core State Standard 5.NF.6, 6.NS.1 and 7.NS.2.

Activity IA12.4

Reciprocals

Intended learning: Students will learn about a quantitative meaning for a reciprocal and reflect on patterns in their reasoning.

Instructional mode: Students working in pairs with initial instruction from the teacher.

Materials: Paper with a variety of different-sized rectangles on it, pencils, markers.

Description: In this activity, students work on developing a quantitative meaning for a reciprocal by working on problems like the 'One-fourth of a Metre Problem' and the 'Two-third of a Mile Problem'. Give each pair a fraction of a metre that is represented by a rectangle on their paper. They are to

(1) draw the whole metre; (2) determine how many times their fraction of a metre fits into a whole metre; and (3) determine what to multiply their fraction by to make 1 metre. In whole-class discussion students can present their work to the class. Following the experience of many problems, ask students to reflect on the patterns in their reasoning, as in the 'Patterns in Our Reasoning Problem'.

Responses, Variations and Extensions

* This activity can be done in *JavaBars*.
* As students gain facility with the activity, the student in each pair could have the same fraction but a different-size bar. Class discussion could address that the size of the bar will produce different-sized parts, but it will not change the answers to parts (2) and (3) of the question. Why?

Activity IA12.5

Solving Measurement Division Problems II

Intended learning: Students will learn how to make multiplicative solutions for measurement division problems involving fractions.

Instructional mode: Students working in pairs or individually with initial instruction from the teacher.

Materials: Paper, writing utensils, worksheet of problems like the 'Flowerbed Problem, Take 2'.

Description: In this activity, students work on developing multiplicative solutions to fraction measurement division problems, as laid out in the four parts of the 'Flowerbed Problem, Take 2'. We recommend using contexts that involve mass, weight and capacity (as in the 'Flowerbed Problem, Take 2' and the problems in IA12.3). Here we list some good number combinations to start with and some good number combinations that are more challenging. In all cases, if students develop a multiplicative solution, it is helpful to ask how they know they can multiply to solve the problem.

Some number combinations to start with:

$5 \div \frac{1}{4}$, or any whole number divided by a unit fraction

$5 \div \frac{2}{3}$, or any whole number divided by a fraction consisting of two fractional parts (e.g. $\frac{2}{3}, \frac{2}{5}, \frac{2}{7}$)

$5 \div \frac{3}{4}$, or any whole number divided by $\frac{3}{4}$

$5\frac{2}{3} \div \frac{1}{6}$, or any mixed number divided by a unit fraction, where the proper fraction in the mixed number is a whole number multiple of the unit fraction

$5\frac{2}{3} \div \frac{4}{3}$, or any mixed number divided by an improper fraction, where the improper fraction is a whole number multiple of the proper fraction in the mixed number

Some number combinations that are more challenging:

$5\frac{1}{2} \div \frac{1}{3}$, or any whole number plus $\frac{1}{2}$, divided by a unit fraction, where $\frac{1}{2}$ is not a whole number multiple of the unit fraction (i.e. $5\frac{1}{2} \div \frac{1}{4}$ is easier than $5\frac{1}{2} \div \frac{1}{3}$)

(Continued)

(Continued)

$5\frac{2}{3} \div \frac{1}{4}$, or a mixed number divided by a unit fraction where the proper fraction in the mixed number is not a whole number multiple of the unit fraction

$5\frac{2}{3} \div \frac{3}{4}$, or a mixed number divided by a fraction where the proper fraction in the mixed number is not a whole number multiple of the fraction

$5\frac{1}{2} \div \frac{8}{5}$, or a mixed number divided by an improper fraction where the improper fraction is not a whole number multiple of the proper fraction in the mixed number

Responses, Variations and Extensions

- If students have developed fractions as numbers (Chapter 8), improper fractions can be used for the total amount as well as the known share size.

13

From Fractions
to Algebra

The push to engage all students in algebraic reasoning (e.g. NCTM, 2000) has led to the study of how children's ideas about algebra can grow out of their ideas about number and quantity (e.g. Carpenter et al. 2003; Kaput et al., 2008; Russell et al., 2011). Although the focus has often been on connections between students' whole number knowledge and algebraic reasoning, scholars have begun to understand the connections between students' fractions knowledge and algebraic reasoning (e.g. Ellis, 2007; Empson et al., 2011; Hackenberg and Lee, 2015). In this chapter we outline a few of these ideas, and address how students at Stages 2 and 3 might work on the other meaning of fraction division, which is a version of the partitive, or sharing, meaning.

Fractions and Algebra: Similar Ways of Thinking

Following the work of Kaput (2008), we view students' beginning algebraic reasoning to be about:

(1) **generalizing** and abstracting arithmetical and quantitative relationships, and systematically representing those generalizations and abstractions, not necessarily with standard algebraic notation, and

(2) learning to reason with algebraic notation as a stand-in for reasoning with quantities.

The view that algebraic reasoning includes aspect (1) above means that students can be reasoning algebraically well before they are using standard algebraic notation, such as symbolizing unknown or varying quantities with letters. It also means that when students make generalizations about whole numbers and fractions, those generalizations can be seen to have an algebraic character that teachers can help students to articulate and record with drawings, words and numerical expressions and sentences. Systematically articulating in this way is part of what makes students' generalizations algebraic.

For example, a 5th-grade student named Trenton was solving a measurement division problem where a frog ate $1\frac{1}{2}$ cups of food each day. The problem asked for how many days the frog could eat if his owner has 12 cups of food. Trenton reasoned that 2 times $1\frac{1}{2}$ was 3 'because 2×1 is 2 and $2 \times \frac{1}{2}$ is 1, and then $2 + 1$ is 3' (Empson and Levi, 2011: 60). So, he knew that the frog would need 3 cups of food to eat for 2 days. Then he saw the 12 cups as 4 times the 3 cups, and he reasoned that the frog could eat for 4 times as many days as 3 cups would feed him, so 4×2 days, or 8 days.

Empson and Levi (2011) identify that Trenton implicitly used fundamental mathematical properties in his thinking. An implicit use of a property means that mathematics teachers can see in Trenton's reasoning the use of mathematical properties, even though Trenton himself was probably not aware of using them. For example, in multiplying $2 \times 1\frac{1}{2}$, Trenton multiplied 2×1 and $2 \times \frac{1}{2}$, which is a use of the Distributive Property of Multiplication over Addition. If we were to record Trenton's thinking to show this property, we might write as follows: $2 \times 1\frac{1}{2} = 2(1 + \frac{1}{2}) = 2 \times 1 + 2 \times \frac{1}{2} = 2 + 1 = 3$.

Empson and Levi (2011) call this kind of reasoning *relational thinking* with fractions. They identify it as a connection between fractions and algebra because it represents reasoning with structures of our number systems – and representing and using these structures is a part of algebraic thinking. Students who regularly engage in relational thinking with fractions and whole numbers, and who are asked to articulate and record their relational thinking, are learning algebraic uses of fundamental mathematical properties. So, Trenton might come to see that $\frac{1}{2}(a + b) = \frac{1}{2}a + \frac{1}{2}b$ because the underlying property is the same as the one he used to multiply $2 \times 1\frac{1}{2}$.

In addition, recent research has suggested that the thinking that goes into conceiving of fractions as numbers in the way we addressed in Chapter 8, which involves the coordination of three levels of units, may be very useful for reasoning with quantitative unknowns and variables (Hackenberg and Lee, 2015). That is, conceiving of fractions as numbers and reasoning with quantitative unknowns and variables may involve similar ways of mentally structuring units. We give an example of this in the following two sections. Incidentally, this finding means the link could go both ways: Reasoning with quantitative unknowns and variables might support students in advancing their fractions knowledge.

Quantitative Unknowns

For us, *quantities* are measurable properties of a person's concept of an object or phenomenon (Thompson, 2010). A *measurable* property involves a measurement unit and a measurement process. For example, height is a measurable property of our concept of a person. The property, height, is a distance or span from the soles of a person's feet to the top of their head. We can talk about a person's height without knowing how tall that person is. However, even when we do not know a value for a quantity, we can still imagine a measurement unit and measurement process that could be used to find a value for a person's height, such as subdividing the height into centimetres (a measurement unit) and counting them up (a measurement process).

A quantitative perspective on algebra means that an *unknown* is a value of a quantity that we currently do not know, but could find out. So, an unknown is a potential measurement of a quantity. Figure 13.1 is a picture that could represent a known length – its value is 14 units. Figure 13.2 is a picture that would represent an unknown length – its value is not known but we can still imagine partitioning up the length into equal parts and, say, counting them to find a value.

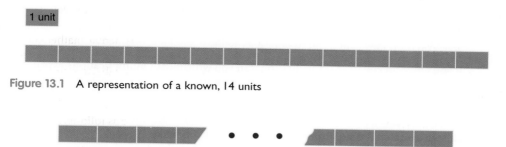

Figure 13.1   A representation of a known, 14 units

Figure 13.2   A representation of an unknown

As people move toward learning to reason with algebraic notation as a stand-in for reasoning with quantities, our second aspect of beginning algebraic reasoning (see aspect (2) above), instead of the image in Figure 13.2 we might use a letter, say *q*, to represent the unknown length. But for many students it is helpful to refer back to a quantitative image for some time, even after using standard algebraic notation.

Multiplicative Reasoning with Quantitative Unknowns

One characteristic of quantitative unknowns as we have defined them above is that they require thinking about a quantity – let's take a basic quantity, like a length – as partitioned into some number of equal parts, where the number of parts is unspecified. As we have discussed in Chapters 3 and 4, that is precisely what students at Stage 1 cannot yet do. So, this means that students at Stage 1 will be quite challenged to work with quantitative unknowns, and they would likely benefit more from a focus on helping them develop their whole number and fractions knowledge.

In contrast, students at Stages 2 and 3 can take a partitioned length as a given in their thinking prior to working in a situation. So, they can conceive of quantitative unknowns. However, how they think about quantitative unknowns as they work further with them in solving problems has been found to differ (Hackenberg and Lee, 2015). Specifically, students at Stage 2 have been found to develop what we call inverse reasoning with quantitative unknowns, while students at Stage 3 have been found to develop what we call reciprocal reasoning with quantitative unknowns (Hackenberg et al., 2015; Hackenberg and Sevis, 2015).

Students at Stages 2 and 3 can work on a problem like the following:

'Corn Stalk Tomato Plant Heights Problem': A corn stalk and a tomato plant are growing in the garden, each of unknown height. The height of the corn stalk measured in inches is five times the height of the tomato plant measured in inches.

a)   Draw a picture of this situation and describe what your picture represents.

b)   Write an equation for this situation that relates the two heights. Explain what your equation means in terms of your picture.

c)   Can you write another, different equation that relates the two heights? Explain what your equation means in terms of your picture.

d)   If you wrote an equation using division, can you write it with multiplication? Explain what your new equation means in terms of your picture.

e)   Let's say that the corn stalk's height is 150 cm. How tall is the tomato plant? Use this example to check all of your equations. If an equation does not work, see if you can change it so that it does. Explain any changes that you make.

We have found that students at Stage 2 often view this problem as involving a taller and shorter height, but they do not necessarily see or represent the taller height as precisely five times the length of the shorter height (see, for example, Figure 13.3).

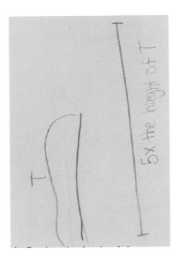

Figure 13.3   A student's initial drawing of the two unknown heights in the 'Corn Stalk Tomato Plant Heights Problem'

However, students can learn to make a more precise relationship – and they often do so by drawing a segment to represent the shorter height, copying it and iterating it to make the taller height as five segments equal to the shorter height (Figure 13.4).

tomato   corn stalk

Figure 13.4   A student's revised drawing of the two unknown heights in the 'Corn Stalk Tomato Plant Heights Problem'

Students at Stage 2 also can learn to write equations that involve multiplication and division for the two heights. However, they do not necessarily write multiplicative equations at the beginning. For example, they might write $m = 4 + p$, where m represents the corn stalk height in inches and p represents the tomato plant height in inches. This equation comes from thinking about adding four more parts onto the tomato plant height to make the corn stalk height. We have found many students at Stage 2 can learn to change their equations to involve multiplication. Sometimes this comes from writing $m = p + p + p + p + p$, and then shortening this to $m = p \times 5$ or $m = 5p$. They can usually articulate this equation as meaning that five tomato plant heights make up a corn stalk height, and they can see that in their pictures. These students can also learn to write $m \div 5 = p$ or $m \div p = 5$, but learning how to see those equations in their pictures of the quantities requires significant effort (Hackenberg et al., 2015). They are very unlikely to write an equation like $\frac{1}{5} m = p$ or to think it is possible to multiply a taller height by any number to produce a shorter height (Hackenberg, 2010).

Furthermore, consider changing the problem so there is a fractional relationship: for example, the tomato plant height in inches is $\frac{1}{4}$ the corn stalk height in inches. In this case, students at Stage 2 may not use the given fraction in their equations at all. For this problem with $\frac{1}{4}$ as the relationship, one 7th-grade student wrote for his equation $\frac{Q}{Z} = ?$, where Q was the corn stalk height in inches and Z was the tomato plant height in inches (Hackenberg et al., 2015). He told the first author that he 'didn't have a number to go on', which was the reason for using the question mark. When the first author questioned him about how to use the smaller height to make the larger height, he made several suggestions, such as using exponents (he was studying exponents in his mathematics class). Then he looked at his picture and said 'you add Z four times to make Q', writing $Z + Z + Z + Z = Q$. This led him to produce both $4Z = Q$ and $Q \div 4 = Z$ for his equations. He and other students at Stage 2 articulated that these equations were sensible because (whole number) multiplication and division were inverses, so we call this inverse reasoning with quantitative unknowns. However, writing an equation involving $\frac{1}{4}$ was an idea he and other students at Stage 2 never generated or found sensible.

In contrast, students at Stage 3 can work with any fractional relationship between the two quantitative unknowns, and they can produce reciprocal reasoning with quantitative unknowns (Hackenberg and Sevis, 2015). So, let's say that the tomato plant height in inches was $\frac{3}{5}$ of the corn stalk height in inches. Students at Stage 3 can learn to generate the equation $s = \frac{3}{5}f$, where s represents the tomato plant height in inches and f represents the corn stalk height in inches. Based on their pictures (see, example, Figure 13.5), they can learn to see the corn stalk height as five equal parts, each of which is $\frac{1}{3}$ of the tomato plant height. So, the corn stalk height is $\frac{5}{3}$ of the tomato plant height. Students at Stage 3 can learn to write $f = \frac{5}{3}s$, or similar equations (e.g. $f = s + \frac{2}{3}s$).

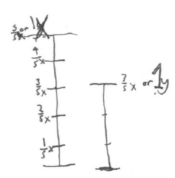

Figure 13.5 A student's drawing of the 'Corn Stalk Tomato Plant Heights Problem' with a $\frac{3}{5}$ relationship

A main reason that students at Stage 3 can create the reciprocal relationship is because they have developed fractions as numbers (Hackenberg and Sevis, 2015). That is, the three-levels-of-units coordinations required to produce fractions as numbers, as we have laid out in Chapter 8, is a useful resource in reasoning with quantitative unknowns that are multiplicatively related because there are many coordinations to make. In the example shown in Figure 13.5, students have to think of the tomato plant height as a unit of three units each containing an unspecified number of units, and then use that three-levels-of-units structure to see the corn stalk height as a unit of five of those units, each containing an unspecified number of units. These coordinations are very similar to those that students make in developing fractions as numbers. Without those coordinations, students are challenged to develop reciprocal reasoning.

The 'Other' Fraction Division

Students at Stage 3 are poised to develop the other meaning for fraction division, which is a version of the partitive or sharing meaning of division (Ma, 1999; Siebert, 2002). In the sharing meaning of whole number division, a known quantity is to be shared into a known number of equal shares and the size of those shares is not known. This is sensible if both knowns are whole numbers, even if the size of the share will be a fraction. For example:

> 'Sticker Problem': Twenty-eight stickers are to be shared equally into seven party bags. How many stickers can be put in each bag?

> 'Sub Sandwich Problem'. Three sub sandwiches are to be shared equally by seven people. How much can each person get?

(We note that we address student work on problems like the 'Sub Sandwich Problem' in Chapter 9.) However, the sharing meaning of division does not seem to be sensible if the number of equal shares is a fraction, because it seems odd to share something into a fractional number of groups.

As has been written about elsewhere (e.g. Siebert, 2002), a way to think about this meaning of division with fractions is to think about the known quantity as representing not just a whole number of shares, but a fractional amount of one share. So, for example, three sub sandwiches could be $\frac{1}{2}$ of the share for one person; how much does one person get? Or, three sub sandwiches could be $2\frac{1}{4}$ the size of the share for one person; how much does one person get?

We think that there are (at least) two adjustments in thinking required here. First, students need to reinterpret whole number division problems as a known quantity that is so many whole number of times the size of one share. For example:

> 'Sticker Problem, Take 2': Twenty-eight stickers is 7 times the amount needed for one party bag; how many stickers go in one bag?

We note that reasoning to solve the 'Sticker Problem, Take 2' requires a splitting operation (see Chapter 7). So that puts this first adjustment in the realm of Stage 2 and Stage 3 students.

Second, students have to think that it makes sense to have the known quantity represent a fractional amount of one share. That is, we can change the 7 times to be any fraction or whole number we want. In general, this will only make sense if the known quantity can be subdivided, so we have to abandon stickers and move to objects like sub sandwiches, candy bars, etc. For example:

> 'Sub Problem, Take 2': Three sub sandwiches is $\frac{4}{5}$ the amount for one share; how much is one share?

We call this meaning of fraction division the *algebraic* meaning (cf. Siebert, 2002) because it is closely related to thinking about a division relationship in the 'Corn Stalk Tomato Plant Heights

Problem' when one height is expressed as a fraction of the other height. For example, if the tomato plant height measured in inches is $\frac{1}{4}$ of the corn stalk height measured in inches, then writing $Z \div \frac{1}{4} = Q$ means that Z is $\frac{1}{4}$ the amount of inches needed to make Q. Similarly, if the tomato plant height measured in inches is $\frac{3}{5}$ of the corn stalk height measured in inches, then writing $s \div \frac{3}{5} = f$ means s is $\frac{3}{5}$ of the amount of inches needed to make f. (In contrast, $f \div s = \frac{5}{3}$ highlights a measurement meaning of division, where s fits into f $\frac{5}{3}$ times. Similarly, $s \div f = \frac{3}{5}$ means f fits into s $\frac{3}{5}$ of a time.)

Working on the Algebraic Meaning of Fraction Division

We expect that students at Stage 3 can make progress developing the algebraic meaning for fraction division because it relies on conceiving of fractions as numbers and using fractions multiplicatively.

First, we expect it will be helpful to think about problems like these:

'1-foot Fruit Bar Problem': This rectangle represents a fruit bar. Let's pretend it's 1 foot long.

a) This fruit bar is $\frac{2}{3}$ the size of another bar. Draw the other bar and determine how long it is.

b) Describe your process in part (a).

c) Express your process to make the other bar with division and multiplication of whole numbers.

d) If you divide a length by 2 and then multiply the result by 3, what fraction are you multiplying the length by? How do you know?

Although some students at Stage 2 can solve parts (a) through (c), we think they will find part (d) quite foreign; in fact, part (d) can be challenging for students at Stage 3 as well (Hackenberg, 2010). We recommend doing several problems like the '1-foot Fruit Bar Problem', where 1 unit of length is used, and several different fractions (e.g. 1 yard is $\frac{3}{5}$ the size of another length, 1 metre is $\frac{5}{8}$ the size of another length, etc.).

Following work on these '1-unit' problems, a next step is to change the known quantity to be multiple units. For example:

'Landscape Architect Problem': A landscape architect bought 3 tonnes of soil (Figure 13.6), which was $\frac{2}{3}$ of the amount of soil she needed for her project. How much soil does she need for her project?

a) Draw the amount of soil the architect needs and determine how many tonnes are required.

b) Describe your process in part (a).

c) Express your process to make the other amount of soil with division and multiplication of whole numbers.

d) Express your process to make the other amount of soil with multiplication only.

e) Write a statement of division for this problem. Explain how the two statements in parts (d) and (e) are related.

Figure 13.6   Three tonnes of soil

The given amount of soil can change here to be 10 tonnes, 55 tonnes, or $13\frac{1}{3}$ tonnes. One main issue to be developed is that no matter the given amount of soil, if it is $\frac{2}{3}$ of the amount needed then we have to partition it into two equal parts and take one of those parts three times in order to make the needed amount of soil. Students at Stage 3 may be able to make a general statement like this one, as well as to see that taking one-half of something three times is taking $\frac{3}{2}$ of it. They have the potential to see that because they have developed fractions as numbers (see Chapter 8).

Another main issue to be developed is how to take $\frac{3}{2}$ of a quantity so that one knows how much it is. For example, to take $\frac{3}{2}$ of the 3 tonnes students might take $\frac{1}{2}$ of 3, get $1\frac{1}{2}$ tonnes, and multiply that by 3 to get $4\frac{1}{2}$ tonnes. Alternately, depending on the order of students' ideas, they might take $\frac{3}{2}$ of each tonne and get $\frac{3}{2}$ tonnes times 3. This activity to produce the needed amount of soil requires the ways of thinking students have available at Stage 3. When the numbers involved are more challenging (e.g. $8\frac{1}{2}$ tonnes is $\frac{3}{5}$ of the amount of soil needed), this issue only becomes more evident (Hackenberg, 2010).

We note that just as with measurement division problems, there is a nice connection to ratio reasoning here that we suggest is fruitful to develop with students. For example, we could view solving the 'Landscape Architect Problem' as a process of creating equal ratios, such as (3 tonnes)/($\frac{2}{3}$ of what she needs) = ($4\frac{1}{2}$ tonnes)/(1 whole amount of what she needs). So, problems like these can be used to interpret the given fractional relationship as a ratio between two quantities, the part of the soil that she needs and the whole amount. And there could be many possible amounts that would yield that same ratio, depending on the size of the amount the landscape architect has started with. We believe this way of thinking can shed even more light on the nature of division and generalizations of it.

Glossary

Associative property of multiplication. The property of multiplication guaranteeing that $a \times b \times c = (a \times b) \times c = a \times (b \times c)$.

Associative reasoning. Problem-solving strategies that take advantage of an implicit understanding of the Associative Property of Multiplication.

Bricolage. A means of problem-solving that involves tinkering with a diverse set of immediately available items (see Chapter 9).

Commensurate fractions. Two or more fractions that have the same size, or measure.

Composite unit. A unit made up of smaller units.

Connected number. A sequence of continuous wholes created by iterating a continuous unit so many times (see Chapter 4).

Disembedding. Taking a part out of a whole without destroying the whole (see Chapter 2).

Distributing. Inserting the units within one composite unit into each of the units in another composite unit (see Chapter 2).

Distributive Property (of Multiplication over Addition). The relationship between multiplication and addition in which multiplicities of a sum can be taken over its parts. In other words $a \times (b + c) = (a \times b) + (a \times c)$.

Distributive reasoning. Sharing a collection of items equally into n parts by sharing each of its items equally into n parts (see Chapter 9); taking a fraction of a quantity by taking that fraction of parts of that quantity (see Chapter 9, 10).

Dividend. When dividing two numbers, the dividend is the number that is being divided (by the divisor).

Divisor. When dividing, the divisor is the number that is being used to divide the other number (the dividend).

Equal sharing. Creating same-sized parts from a whole (a continuous whole or a collection of items).

Equi-partitioning. Partitioning a continuous whole into equally-sized pieces, understanding that any one of those pieces could be iterated to reproduce the whole.

Equivalent fractions. An equivalence class of fractions, where all of the elements of the class are represented by a single fraction to which the other fractions simplify.

Fraction number sequence. A set of lengths created by taking multiples of a unit fraction, such as 1/5, 2/5, 3/5, … (see Chapter 8).

Fractional unit. A unit fraction treated as a unit of measure, relative to the whole.

Fraction rods. A set of rectangular prisms whose lengths vary from 1 to 10 (also known as 'Cuisenaire rods'; see illustration in Appendix).

Fraction strips. Long, thin rectangular pieces of paper (see Appendix).

Fragmenting. Breaking a whole into parts (see Chapter 4).

Improper fraction. A fraction whose measure is greater than the whole.

Initial number sequence. The first number sequence that students construct in order to quantify numbers.

Iterating. Making connected copies of a part by repeating it (see Chapter 2).

Iterative Fraction Scheme. A way of conceiving and working with fractions in which non-unit fractions gain meaning as multiples of unit fractions; three levels of units – the fraction, the unit fraction and the whole – are coordinated in conceptualizing fractions and maintained in working with them (see Chapter 8).

Magic candy bar (or cake). An imaginary whole that replenishes itself after a part is taken out. This pedagogical tool is especially useful for instruction on improper fractions.

Measurement Scheme for Unit Fractions. A way of conceptualizing and working with fractions in which the unit fraction, $1/n$, gains its meaning through a 1-to-n relationship with the whole (see Chapter 6).

Mixed number. A number with an integer part and a fractional part.

Multiplicative reasoning. Working with multiples of a composite unit on at least two levels simultaneously: the multiplicity of the composite units and the multiplicity of the units it contains.

Non-unit fraction. A fraction that is not a unit fraction. In other words, its numerator is not 1.

Operation. A mental action (internalized action) that has been organized with other mental actions.

Partitioning. The operation of creating equal parts within a continuous whole (see Chapter 4).

Partitive (or sharing) division. A meaning for division in which a known quantity is to be shared into a known number of equal shares and the size of those shares is not known.

Parts Out of Wholes Fraction Scheme. A way of conceptualizing and working with proper fractions in which the fraction is taken out of (disembedded from) and compared to the whole, by way of the number of equal parts in the fraction and the whole (see Chapter 5).

Parts Within Wholes Fraction Scheme. A way of conceptualizing and working with proper fractions in which the fraction is compared to the whole, by way of the number of equal parts in the fraction and the whole in which the fraction is contained (see Chapter 5).

Proper fraction. A fraction whose measure does not exceed the whole.

Quantitative unknown. A value of a quantity that we currently do not know, but that we could find out – a potential measurement of a quantity (see Chapter 13).

Quotitive (or measurement) division. A measurement meaning for division in which a known quantity is to be shared into shares of a known size, and the number of equal shares is unknown (see Chapter 12).

Reciprocal. Formally the multiplicative inverse of a fraction, a quantitative meaning for the reciprocal refers to the fraction of a fractional quantity needed to produce a whole unit of the quantity (see Chapter 12).

Relational thinking. Reasoning explicitly or implicitly with the structure of one's number systems, such as with mathematical properties (see Chapter 13).

Reorganization hypothesis. The conjecture that students construct their knowledge of fractions by reorganizing their ways of working with whole numbers (see Chapter 2).

Repeated halving. Creating fourths, eighths and other fractional powers of 2 by recursively halving halves (see Chapter 9).

Reversible reasoning. A student's ability to reverse the operations of their fraction scheme to, for example, reproduce the whole from a proper fraction of it (see Chapter 7).

Scheme. A student's way of working with, or conceiving of, a mathematical object, such as a fraction.

Segmenting. Using a smaller length to partition off segments within a larger length.

Simultaneous partitioning. Partitioning a continuous whole into equal parts by projecting a composite unit into it (see Chapter 2).

Splitting. The simultaneous coordination of partitioning and iterating (see Chapter 7).

Unit fraction. Named $1/n$, a unit fraction is a part (or unit) that results from partitioning a whole into n equal parts; conversely, this is the part that can be iterated n times to reproduce the whole.

Unitizing. The mental act of regarding a collection as a unit (see Chapter 2).

Units coordinating. Working with multiple levels of units at the same time (see Chapters 2 and 3).

Whole unit. The unit to which fractional units refer.

Appendix: Templates Marked and Unmarked Fraction Strips

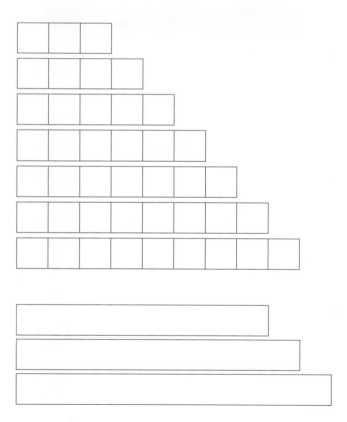

Fraction Rods

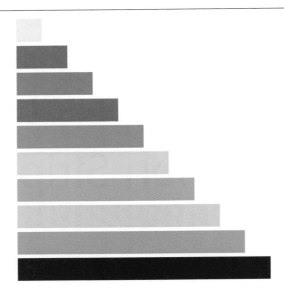

References

Behr, M.J., Harel, G., Post, T.R. and Lesh, R. (1993) Rational numbers: Toward a semantic analysis – emphasis on the operator construct. In T.P. Carpenter, E. Fennema and T.A. Romberg (eds), *Rational Numbers: An Integration of Research*. Hillsdale, NJ: Lawrence Erlbaum Associates, Inc. pp. 13–47.

Behr, M.J., Wachsmuth, I., Post, T.R. and Lesh, R. (1984) Order and equivalence of rational numbers: a clinical teaching experiment. *Journal for Research in Mathematics Education*, 15, 323–41.

Biddlecomb, B.D. (2002) Numerical knowledge as enabling and constraining fraction knowledge: an example of the reorganization hypothesis. *Journal of Mathematical Behavior*, 21, 167–90.

Boyce, S. and Norton, A. (2016) Co-construction of fractions schemes and units coordinating structures. *Journal of Mathematical Behavior*, 41, 10–25

Carpenter, T.P., Franke, M.L. and Levi, L.W. (2003) *Thinking Mathematically: Integrating Arithmetic and Algebra in Elementary School*. Portsmouth, NH: Heinemann.

Common Core State Standards for Mathematics (2010) www.corestandards.org/assets/CCSSI_Math%20Standards.pdf

Ellemor-Collins, D. and Wright, R.J. (2011) Unpacking mathematisation: an experimental framework for arithmetic instruction. In B. Ubuz (ed.), *Proceedings of the 35th Conference of the International Group for the Psychology of Mathematics Education*, vol. 2. Ankara, Turkey: PME. pp. 313–20.

Ellis, A.B. (2007) Connections between generalizing and justifying: students' reasoning with linear relationships. *Journal for Research in Mathematics Education*, 38 (3), 194–229.

Empson, S.B. and Levi, L. (2011) *Extending Children's Mathematics: Fractions and Decimals*. Portsmouth, NH: Heinemann.

Empson, S.B., Levi, L. and Carpenter, T.P. (2011) The algebraic nature of fractions: developing relational thinking in elementary school. In J. Cai and E.J. Knuth (eds), *Early Algebraization*. Berlin: Springer-Verlag. pp. 409–28.

'fraction' (2015) In *Merriam-Webster.com*. Retrieved 17 August 2015 from www.merriam-webster.com/dictionary/fraction

Gould, P. (2012) Connecting the teaching and learning of fractions. In R.J. Wright, D. Ellemor-Collins and P. Tabor (eds), *Developing Number Knowledge: Assessment, Teaching and Intervention with 7–11 year olds*. London: Sage.

Gravemeijer, K.P.E., Cobb, P., Bowers, J.S. and Whitenack, J.W. (2000) Symbolizing, modeling and instructional design. In P. Cobb, E. Yackel and K.J. McClain (eds), *Symbolizing and Communicating in Mathematics Classrooms: Perspectives on Discourse, Tools, and Instructional Design*. Hillsdale, NJ: Lawrence Erlbaum Associates, Inc. pp. 225–73.

Hackenberg, A.J. (2007) Units coordination and the construction of improper fractions: a revision of the splitting hypothesis. *Journal of Mathematical Behavior*, 26, 27–47.

Hackenberg, A.J. (2010) Students' reasoning with reversible multiplicative relationships. *Cognition and Instruction*, 28 (4), 383–432.

Hackenberg, A.J. (2013) The fractional knowledge and algebraic reasoning of students with the first multiplicative concept. *Journal of Mathematical Behavior*, 32 (3), 538–63.

Hackenberg, A.J., Jones, R., Eker, A., Creager, M. and Timmons, R. (2015) *'Approximate' Multiplicative Relationships between Quantitative Unknowns*. Paper presented at the American Educational Research Association (AERA) Conference, Chicago, IL.

Hackenberg, A.J. and Lee, M.Y. (2015) Relationships between students' fractional knowledge and equation writing. *Journal for Research in Mathematics Education*, 46 (2), 196–243.

Hackenberg, A.J. and Lee, M.Y. (in press) Students' distributive reasoning with fractions and unknowns. *Educational Studies in Mathematics*,

Hackenberg, A.J. and Sevis, S. (2015) *On a Learning Trajectory for Reciprocal Reasoning with Quantitative Unknowns*. Paper presented at the American Educational Research Association (AERA) Conference, Chicago, IL.

Hackenberg, A.J. and Tillema, E.S. (2009) Students' whole number multiplicative concepts: a critical constructive resource for fraction composition schemes. *Journal of Mathematical Behavior*, 28, 1–18.

Hunting, R.P. and Davis, G.E. (1991) Dimensions of young children's conceptions of the fraction one-half. In R.P. Hunting and G. Davis (eds), *Early Fraction Learning*. New York: Springer. pp. 27–53.

Kaput, J.J. (2008) What is algebra? What is algebraic reasoning? In J.J. Kaput, D.W. Carraher and M.L. Blanton (eds), *Algebra in the Early Grades*. New York: Lawrence Erlbaum. pp. 5–17.

Kaput, J.J., Carraher, D.W. and Blanton, M.L. (eds) (2008) *Algebra in the Early Grades*. New York: Lawrence Erlbaum.

Kieren, T.E. (1980) The rational number construct – its elements and mechanisms. In T.E. Kieren (ed.), *Recent Research on Number Learning*. Columbus: ERIC/SMEAC. pp. 125–49.

Lamon, S.J. (1996) The development of unitizing: its role in children's partitioning strategies. *Journal for Research in Mathematics Education*, 27(2), 170–93.

Lee, M.Y. and Aydeniz, F. (2015) *Investigating a Student's Distributive Partitioning Scheme: The Case of Gabriel*. Paper presented at the American Educational Research Association (AERA) Conference, Chicago, IL.

Li, Y., Chen, X. and An, S. (2009) Conceptualizing and organizing content for teaching and learning in selected Chinese, Japanese, and U.S. mathematics textbooks: the case of fraction division. *ZDM*, 41, 809–26.

Ma, L. (1999) *Knowing and Teaching Elementary Mathematics: Teachers' Understanding of Fundamental Mathematics in China and the United States*. Mahwah, NJ: Lawrence Erlbaum Associates.

NCTM (2000) *Principles and Standards for School Mathematics*. Reston, VA: National Council of Teachers of Mathematics.

Norton, A. (2008) Josh's operational conjectures: abductions of a splitting operation and the construction of new fractional schemes. *Journal for Research in Mathematics Education*, 39 (4), 401–30.

Norton, A., and Boyce, S. (2013) A cognitive core for common state standards. *Journal of Mathematical Behavior*, 32 (2), 266–79.

Norton, A., and Boyce, S. (2015) Provoking the construction of a structure for coordinating $n+1$ levels of units. *Journal of Mathematical Behavior*, 40, 211–42. doi:10.1016/j.jmathb.2015.10.006

Norton, A., Boyce, S. and Hatch, J. (2015) Coordinating units at the Candy Depot. *Mathematics Teaching in the Middle School*, 21 (5), 280–7.

Norton, A., Boyce, S., Phillips, N., Anwyll, T., Ulrich, C. and Wilkins, J. (2015) A written instrument for assessing students' units coordination structures. *Journal of Mathematics Education*, 10 (2), 111–36.

Norton, A. and Wilkins, J.L.M. (2013) Supporting students' constructions of the splitting operation. *Cognition & Instruction*, 31 (1), 2–28.

Norton, A., Wilkins, J.L.M., Evans, M.A., Deater-Deckard, K., Balci, O. and Chang, M. (2014) Technology helps students transcend part–whole concepts. *Mathematics Teaching in the Middle School*, 19 (6), 352–9.

Olive, J. and Vomvoridi, E. (2006) Making sense of instruction on fractions when a student lacks necessary fractional schemes: the case of Tim. *Journal of Mathematical Behavior*, 25, 18–45.

Piaget, J., Inhelder, B. and Szeminska, A. (1960) *The Child's Conception of Geometry*. New York: Routledge.

Russell, S.J., Schifter, D. and Bastable, V. (2011) *Connecting Arithmetic to Algebra: Strategies for Building Algebraic Thinking in the Elementary Grades*. Portsmouth, NH: Heinemann.

Siebert, D. (2002) Connecting informal thinking and algorithms: the case of division of fractions. In B. Litwiller and G.W. Bright (eds), *Making Sense of Fractions, Ratios, and Proportions*. Reston, VA: National Council of Teachers of Mathematics. pp. 247–56.

Steffe, L.P. (1991) The constructivist teaching experiment: illustrations and implications. In E. von Glasersfeld (ed.), *Radical Constructivism in Mathematics Education*. Boston, MA: Kluwer Academic Press. pp. 177–94.

Steffe, L.P. (1992) Schemes of action and operation involving composite units. *Learning and Individual Differences*, 4 (3), 259–309.

Steffe, L.P. (1994) Children's construction of meaning for arithmetical words: a curriculum problem. In D. Tirosh (ed.), *Implicit and Explicit Knowledge: An Educational Approach*. Norwood, NJ: Ablex. pp. 131–68.

Steffe, L.P. (2002) A new hypothesis concerning children's fractional knowledge. *Journal of Mathematical Behavior*, 20 (3), 267–307.

Steffe, L.P. (2004) On the construction of learning trajectories of children: the case of commensurate fractions. *Mathematical Thinking and Learning*, 6 (2), 129–62.

Steffe, L.P. (2010a) Articulation of the reorganization hypothesis. In L.P. Steffe and J. Olive (eds), *Children's Fractional Knowledge*. New York: Springer. pp. 49–74.

Steffe, L.P. (2010b) Operations that produce numerical counting schemes. In L.P. Steffe and J. Olive (eds), *Children's Fractional Knowledge*. New York: Springer. pp. 27–47.

Steffe, L.P. and Olive, J. (2010) *Children's Fractional Knowledge*. New York: Springer.

Streefland, L. (1991) *Fractions in Realistic Mathematics Education*. Dordrecht, The Netherlands: Kluwer Academic.

Thompson, P. W. (2010) Quantitative reasoning and mathematical modeling. In S. Chamberlin and L. L. Hatfield (eds), *New Perspectives and Directions for Collaborative Research in Mathematics Education*. Laramie, WY: University of Wyoming. pp. 33–57.

Thompson, P.W. and Saldanha, L.A. (2003) Fractions and multiplicative reasoning. In J. Kilpatrick, W.G. Martin and D. Schifter (eds), *A Research Companion to Principles and Standards for School Mathematics*. Reston, VA: National Council of Teachers of Mathematics. pp. 95–113.

Tillema, E.S. and Hackenberg, A.J. (2011) Developing systems of notation as a trace of reasoning. *For the Learning of Mathematics*, 31 (3), 29–35.

Turkle, S. and Papert, S. (1992) Epistemological pluralism and the revaluation of the concrete. *Journal of Mathematical Behavior*, 11, 3–33.

Tzur, R. (1999) An integrated study of children's construction of improper fractions and the teacher's role in promoting that learning. *Journal for Research in Mathematics Education*, 30 (4), 390–416.

Vygotsky, L.S. (1978) *Mind in Society: The Development of Higher Psychological Processes*. Cambridge, MA: Harvard University Press. (Translator M. Lopez-Morillas, original work published 1934.)

Watanabe, T. (2007) Initial treatment of fractions in Japanese textbooks. *Focus on Learning Problems in Mathematics*, 29 (2), 41–60.

Wilkins, J. and Norton, A. (2011) The splitting loope. *Journal for Research in Mathematics Education*, 42 (4), 386–406.

Wright, R.J. (1994) A study of the numerical development of 5-year-olds and 6-year-olds. *Educational Studies in Mathematics*, 26, 25–44.

Wright, R.J. (2000) Professional development in recovery education. In L.P. Steffe and P.W. Thompson (eds), *Radical Constructivism in Action: Building on the Pioneering Work of Ernst von Glasersfeld*. London: Falmer. pp. 134–51.

Wright, R.J. (2003) Mathematics Recovery: a program of intervention in early number learning. *Australian Journal of Learning Disabilities*, 8 (4), 6–11.

Wright, R.J. (2008) Mathematics Recovery: an early intervention program focusing on intensive intervention. In A. Dowker (ed.), *Mathematics Difficulties: Psychology and Intervention*. San Diego, CA: Elsevier. pp. 203–23.

Wright, R.J. (2013) Assessing early numeracy: significance, trends, nomenclature, context, key topics, learning framework and assessment tasks. *South African Journal of Childhood Education*, 2, 21–40.

Wright, R.J., Ellemor-Collins, D. and Tabor, P. (2012) *Developing Number Knowledge: Assessment, Teaching and Intervention with 7–11 year olds*. London: Sage.

Wright, R.J., Martland, J. and Stafford, A. (2006a) *Early Numeracy: Assessment for Teaching and Intervention*, 2nd edn. London: Sage.

Wright, R.J., Martland, J., Stafford, A.K. and Stanger, G. (2006b) *Teaching Number: Advancing Children's Skills and Strategies*, 2nd edn. London: Paul Chapman.

Wright, R.J., Stanger, G., Stafford, A. and Martland, J. (2015) *Teaching Number in the Classroom with 4–8 year olds*, 2nd edn. London: Sage.

Yang, D., Reys, R.T. and Wu, L. (2010) Comparing the development of fractions in the fifth- and sixth-graders' textbooks of Singapore, Taiwan, and the USA. *School Science and Mathematics*, 110 (3), 118–27.

Index

Tables are indicated by page numbers in bold print.